Discovery or Construction?

Philosophie und Geschichte der Wissenschaften

Studien und Quellen

Herausgegeben von
Hans Jörg Sandkühler (Bremen) und
Pirmin Stekeler-Weithofer (Leipzig)

Band 73

PETER LANG
Frankfurt am Main · Berlin · Bern · Bruxelles · New York · Oxford · Wien

Francisco José Soler Gil

Discovery or Construction?

Astroparticle Physics and the
Search for Physical Reality

PETER LANG
Internationaler Verlag der Wissenschaften

Bibliographic Information published by the Deutsche Nationalbibliothek
The Deutsche Nationalbibliothek lists this publication in the Deutsche Nationalbibliografie; detailed bibliographic data is available in the internet at http://dnb.d-nb.de.

ISSN 0724-4479
ISBN 978-3-631-63720-3
© Peter Lang GmbH
Internationaler Verlag der Wissenschaften
Frankfurt am Main 2012
All rights reserved.

All parts of this publication are protected by copyright. Any utilisation outside the strict limits of the copyright law, without the permission of the publisher, is forbidden and liable to prosecution. This applies in particular to reproductions, translations, microfilming, and storage and processing in electronic retrieval systems.

www.peterlang.de

To my colleagues and friends of the research groups
«Astroteilchenphysik» and «theoretische Philosophie» at
the Technische Universität Dortmund, in gratitude for an
unforgettable research stay

Contents

Acknowledgments		11
Introduction		13
1.	The Situation after the Science Wars	13
2.	The Aim of this Study	13
3.	The Importance of the Debate between Realism and Social Constructivism about the Dynamics of Research	15
4.	The Structure of this Investigation	18
1. Chapter: The Realism Debate		21
1.	Introduction	21
2.	The Realism Debate in the History of Physics	22
3.	The Social Constructivist View on the Development of Physics	31
4.	The Meaning of the Dispute between Realism and Constructivism	41
5.	Realistic and Constructivist Case Studies	47
6.	Impulses for the Discussion between Realism and Social Constructivism from Astroparticle Physics	51
2. Chapter: Astroparticle Physics: An Overview for Philosophers		55
1.	Introduction	55
2.	Phenomena of Astroparticle Physics	56
2.1	Phenomenology of Charged Cosmic Rays	57
2.2	Phenomenology of Gamma-Rays	59
2.3	Phenomenology of the cosmic neutrinos	67
2.4	Parenthesis for Philosophers: The Concept of «phenomenon»	69
3.	Experiments, Observatories and Techniques in Astroparticle Physics	71
3.1	The Satellite Detectors	72
3.1.1.	PAMELA [Payload for Antimatter Matter Exploration and Light-Nuclei Astrophysics]	73
3.1.2	FERMI-LAT [Fermi Large Area Telescope]	76
3.2	The Balloon-Experiments	79
3.2.1	ATIC [Advanced Thin Ionization Calorimeter]	80
3.2.2	TIGRE [Tracking and Imaging Gamma-Ray Experiment].	81
3.3	The Fluorescence Experiments	82
3.4	The Air Shower Ground Experiments	84
3.4.1	The «Pierre Auger Experiment»	85
3.5	The Image Generating Air-Cherenkov Telescopes	88
3.6	The Underground Detectors	89
3.7	The Experiments of Astroparticle Physics and the Discussion between Realism and Social Constructivism	91

4.	Theoretical Models in Astroparticle Physics	95
4.1	Models of Sources and Spectra of Cosmic Radiation	96
4.2	Models of Dark (and/or Exotic) Matter	103
4.3	Models of Quantum Gravity	106
4.4	The Theoretical Models in Astroparticle Physics and the Debate between Realism and Constructivism	107
5.	Final Comments on the Discovery and Construction of Phenomena in the Field of Astroparticle Physics	109

3. Chapter: MAGIC — 117

1.	Introduction	117
2.	The MAGIC Collaboration	118
3.	The MAGIC Telescopes	120
3.1	Components of the MAGIC telescopes	121
3.2	Tests of MAGIC	130
3.2.1	Tests to Check the Mirrors of MAGIC II	130
3.2.2	Checks to Test the Operability of the Lasers for Optical Transmission of Signals from the Camera to the «Counting House».	133
3.3.3	Tests related to the update of the «Data Acquisition System» of MAGIC I	135
3.3.4	Tests in Relation to the Update of the Camera	136
3.3.5	Testing the Proper Function of the Telescope	138
4.	The Analysis Chain of MAGIC	147
4.1	The Würzburg-Analysis chain	147
3.	Step of Würzburg Chain: GANYMED	151
4.	Step of Würzburg Chain: SPONDE	153
4.2	The Barcelona Analysis Chain	153
4.3	Monte-Carlo Simulations	155
4.3.1	CORSIKA	157
4.3.2	REFLECTOR	157
4.3.3	CAMERA	158
4.4	Philosophical Comments about the Admissibility of the Analysis Method of the MAGIC Data	159
5.	Some Lessons to be drawn from MAGIC regarding the Debate between Realism and Constructivism	161

4. Chapter: IceCube — 165

1.	Introduction	165
2.	«Social Forces»: Extra Scientific Aspects of Large Scientific Collaborations	166
3.	The IceCube Collaboration	170
4.	The Work of the IceCube Collaboration	174
4.1	Introducting Remarks	174

4.2	The IceCube Detector	176
4.3	The Analysis of the IceCube Data	184
4.4	IceCube and AMANDA	190
5.	IceCube and the Debate between Realism and Social Constructivism	193

5. Chapter: Conclusions and Outlook — 201

1.	Introduction	201
2.	Remarks about the Social Constructivist and Realistic Interpretations of Dynamics of Research within the Field of Astroparticle Physics	201
3.	Future Constructivist and Realistic Developments in the Field of Astroparticle Physics	203
4.	Realism or sheer Conservatism? About the interpretation of the Examples Examined in this Study	206
5.	About the Scope of the Results of this Study	211
6.	What Follows when Social Constructivist Analysis of Dynamics of Scientific Research is not Correct?	213

Bibliography — 217

Index — 227

Acknowledgments

This book is the result of a year-long research work as a member of the astroparticle physics research group at Technische Universität Dortmund. This project has only been made possible thanks to Prof. Dr. Dr. Brigitte Falkenburg and Prof. Dr. Dr. Wolfgang Rhode who kindly invited me to participate in their research project «Modellierung und Vereinheitlichung in der Astroteilchenphysik». The present study includes my contribution to this project. The research project «Modellierung und Vereinheitlichung in der Astroteilchenphysik» on (and thus my stay in Dortmund) was funded by the Fritz Thyssen Stiftung.

Both, Prof. Dr. Dr. Brigitte Falkenburg and Prof. Dr. Dr. Wolfgang Rhode, and the members of their respective research groups (for theoretical philosophy and astroparticle physics) at Technische Universität Dortmund positively influenced this study with their suggestions, proposals and criticisms. They have also contributed decisively to correct all sorts of errors in my manuscripts.

I am currently working on a "Ramón y Cajal" research project on the philosophical topics related to current cosmology. One of the central philosophical issues in this field is whether the cosmological models actually describe the real structure of the world, or whether, by contrast, these models are only useful inventions for "saving the phenomena". As particle astrophysics is a key discipline for the investigation of a number of issues that are critical to the current cosmology (such as the nature of dark matter, and the classical or quantum nature of gravity), the consequences for the philosophy of cosmology deriving from the discussion between realism and social constructivism in astroparticle physics are far reaching. For this reason, I could spend a year of my current research project to complete my study of this topic. I thank the Ministerio de Economía y Competitividad of Spain and the University of Seville for supporting this "Ramón y Cajal" project, and funding the present book.

I am indebted to Raphael Bolinger, Claus Beisbart, Niko Naeve, Florian Braun, Reiner Hedrich, Michael Backes, Dominik Neise, Nikola Strah, Malwina Thom, Marlene Doert, Natalie Milke, Tim Ruhe, Jens Dreyer, Martin Shulze, Jan-Hendrik Köhne, Björn Eichmann and Fabian Clevermann. Many thanks for your help!

I also thank the authors of the figures and drawings that illustrate this book for their kind permission to reproduce them here. I hope not having forgotten to mention anyone. But if I did, I apologize for this now.

I am especially indebted to Prof. Allan Franklin for his patient review of a preliminary draft of the manuscript, and for numerous suggestions for improving the book. I could make some of the proposed changes –I hope that all the changes most desperately needed to improve the manuscript have been introduced–, but because the short time frame for publishing the work, a number of suggestions have not be incorporated. I'm sure the book would have been better if I had attended all his comments.

I am responsible for all remaining deficiencies.

Introduction

1. The Situation after the Science Wars

In recent years, the relationship between natural scientists, social scientists and humanists has been becoming calmer again. After the shock of the publication of the first detailed, social constructivist studies of dynamics of scientific research in the eighties of the last century and the storm of criticism that came down onto the natural scientists from many directions (from feminist to post-modern studies) in the nineties and the past decade, it seems as if some form of coexistence between the intellectual parties has been achieved.

Not too long ago, one had considered such a peaceful coexistence between the faculties of natural and social sciences to be very unlikely. Even in 1999, for example, Allan Franklin described the situation as follows:

> «The attitude of many humanists toward science has changed from indifference to distrust, and even hostility. Today science is under attack from many directions, and each of them denies that science provides us with knowledge. Whether it is because of inherent gender bias, Eurocentrism, or the social and career interests of scientists, science is untrustworthy and fatally flawed.»[1]

The mentioned (and complained about!) «attacks» have become more temperate in the meantime. However, the basic attitude of many social scholars in relation to the natural sciences has remained hostile up to today. Only the extravagant episodes of the science wars, which served in part as entertainment for the general public, are missing now: today scientists pursue their research undisturbed, while many philosophers, historians and sociologists explain in their seminars, that this research leads merely to social constructions that are motivated by certain interests, and do not (and cannot) provide knowledge about the real character of nature. Since each specialist speaks for his own public now, and no more interdisciplinary approach is tried, the conflicts can be avoided... at least for now.

2. The Aim of this Study

Therefore, it is time –since the heated controversies at the end of the last century are finally over– to accomplish a new detailed review of the social constructivist analysis of the dynamics of research. Such an examination is the topic of this study. In the following chapters we will primarily deal with the question, whether the phenomena of physics are discovered or rather constructed. For the whole conflict between the realist and the social constructivist concerning their

---

[1] FRANKLIN, A. (1999), *Can that be right?* (Dordrecht, Kluwer 1999) 1.

views on the dynamics of scientific research (at least with regard to physics) is about this issue:

If the physical phenomena are to be understood as constructions for the consolidation of certain theoretical views, then physics is nothing more than a cultural product without any importance for the characterisation of the real way of being of nature. Physics would then, strictly speaking, be a fiction which may be more or less useful, but which is to be regarded, however, as an invented thing in the same sense as literary works or mythologies. If on the contrary the physical phenomena are to be understood as genuine discoveries, then physics supplies a real knowledge about the structure of the world, which can not be e.g. put on the same level as the narrations of the literature.

Since the publication of the first social constructivist studies on the work of physicists some significant contributions to the debate between realism and social constructivism in physics have already been published.[2] The main difference between these contributions and the present study is, that I will analyse this topic –as far as I know for the first time– on the basis of examples from a research field whose development is still completely open. It is the astroparticle physics, a discipline that examines the high-energy particles of cosmic origin.

The selection of astroparticle physics as the field for a comparison between the realistic and the social constructivist explanation of dynamics of research appears to be advantageous for two reasons: first of all, astroparticle physics represents a discipline that is characterized by the use of sophisticated devices and difficult data analysis methods. That is, it offers an ideal environment for the investigation of the question, to what extent the design of the measuring and recording equipment and the details of the data analysis can lead to a certain plasticity of the physical phenomena. One of the most important tasks of the next chapters will therefore be, to present numerous details about the devices and the data analysis in astroparticle physics with the aim to examine whether there actually is free space for the construction of phenomena in this context.

The second reason for the selection of astroparticle physics in our study can be explained as follows:

The case examples from physics analyzed so far by realists and constructivists concerned solely developments (such as discoveries of phenomena or confirmations of theoretical hypotheses) which are already completed. Regarding these examples, the argument always is, that the development which actually took place can be interpreted either better from this or that perspective. However, everyone who is familiar with the philosophical work-methods and topics knows that the problems of interpretation belong to the most difficult tasks of

---

[2] I would like to remark here particularly the works of FALKENBURG (2007), FRANKLIN (1986), FRANKLIN (1999), GALISON (1987), and HACKING (1999).

philosophy. For one can find always new arguments for and against an interpretation of a certain development (or of a particular text or state of affairs). And so the discussion between realism and social constructivism runs into danger –as long as it relates to the interpretation of completed episodes in the history of physics– that the decision for one or the other perspective is regarded as a matter of taste.

With the consideration of a subject as astroparticle physics, whose development is still in full course, however, a new line of discussion opens up. Because the most important episodes in the history of astroparticle physics are not yet complete, and one can argue that these episodes should run differently in the near future, depending on whether the research follow a realistic or a constructivist dynamics.

In the next chapters I, therefore, will address –in addition to the presentation of the key elements of astroparticle physics and the summary of evidence for the evaluation of the realistic and social constructivist positions that can already be obtained by considering the present state of this discipline– especially the task to supply some examples from research developments that should run differently in the near future, depending on whether the astroparticle physicists pursue their research in the sense of realism or in the sense of social constructivism.

This way, I hope that the present study will provide an effective contribution to overcoming the current impasse in the debate between realism and social constructivism.

3. *The Importance of the Debate between Realism and Social Constructivism about the Dynamics of Research*

Before presenting the structure of this investigation it is appropriate to briefly consider the importance of the debate between scientific realism and social constructivism. Even though the question of the discovery or construction of the physical phenomena sounds very specific (a question of detail for specialists) at first, the answer to this question has serious consequences for the role which physics play (and can play) in the context of our western culture.

The fact is that there is currently a more or less latent conflict between the worldview of physics and the most influential currents within humanities. The now prevailing philosophical currents stress the aptitude and the right of individuals to *self-determination* of their own biography. And they conceive the social norms, mechanisms and structures as *social constructions*, which represent both the product of and the enabling condition for the *self-determination* and creative power of the persons. Rational orders and rational structures are the result of *consensus* between persons who give themselves direction and orientation. *Self-determination*, *consensus* and *social construction* are now considered essential categories, with whose assistance the academic elites of our modern society interpret themselves and their world.

But −in stark contrast to this picture−, the physics speaks of a rational, mathematical order of nature that is not the product of our self-determination and not the result of consensus, but exists independently from us and our wishes and decisions. And even more important is the fact that this objective rational order significantly reduces our capacity for self-determination, since it seems very natural to understand ourselves as part of this physical world. The physical order, thus, gives us a nature.

If we consider this contrast between today's humanist and physical way of thinking, then we can understand the true background of the science wars and the deep bitterness and mistrust with which physics are regarded in many cultural circles.

In order not to leave these reflections on a too abstract level −for this could possibly give the impression that the previous comments represent the outline of a rough or even capricious analysis− let us consider a concrete anecdote of the recent conflict between the humanities and the scientific views:

The positively most famous episode of the science wars was Alan Sokal's publication of an article on quantum gravity in the well-known post-modern oriented journal *Social Text*[3]. This article was a hoax, as Sokal revealed shortly afterwards in a further article published in *Lingua Franca*[4]. An angry wave of reactions and counter reactions followed this hoax, with which we do not have to deal here. Very revealing of the current gap between the humanities and the scientific world view are, however, some passages of the article «Anatomy of a Hoax» published by Bruce Robbins (one of the editors of *Social Text*) on this occasion, as well as the answer Sokals to these arguments. For simplicity, let us pay attention only to the following passage from Robbins':

> «Is it in the interests of women, African Americans, and other super-exploited people to insist that truth and identity are social constructions? Yes and no. No, you can't talk about exploitation without respect for empirical evidence and a universal standard of justice. But yes, truth can be another source of oppression. It was not so long ago that scientists gave their full authority to explanations of why women and African Americans (not to speak of gays and lesbians) were inherently inferior or pathological or both. Explanations like these continue to appear in newer and subtler forms. Hence, there is a real need for a social constructionist critique of knowledge.»[5]

And let us compare this argument with Sokal's reply:

---

[3] SOKAL (1996a).
[4] SOKAL, (1996b).
[5] ROBBINS (1996) 58-59.

«What could he mean by that? Is he simply observing that sometimes the truth is bitter? Apparently not, because his very next sentence explains what he means: "It was not so long ago", he says, "that scientists gave their full authority to explanations of why women and African American ... were inherently inferior". But it is Robbins claiming that *that* is truth? I should hope not! Sure, lots of people says things about women and African Americans that are not true; and yes, those falsehoods have sometimes been asserted in the name of "science", "reason" and all the rest. But claiming something doesn't make it true, and the facts that people -including scientists- sometimes make false claims doesn't mean that we should reject or revise the concept of truth.»[6]

What can we learn from this dispute? In order to answer this question, we are to focus on the thesis from the quotation of Robbins' that is most relevant for our discussion: «truth can be another source of oppression». In my opinion this thesis is best understood as follows: if there is an objective order of nature and if the natural sciences reveal this order (in part), then there would be a truth independent from us. If we, however, accept this scenario, couldn't it then be the case that this objective truth diverges from our ideas about the human being, ethics, the social order, etc.? The quoted answer from Sokal clearly misses the point. Because this is really not a question of whether the specific examples mentioned by Robbins reveal a conflict between the truth about the human being and our plans for a better society in. It is a matter of principle: the natural order –if any such order truly exists and if it is recognizable for us– could contradict our creative willpower in this or that point. Thus, for us the truth may be a source of oppression.

This danger can be addressed, however, by accepting the thesis that sciences do not provide any description of an objective order of nature, but merely social constructions, which should not be given a special status in relation to other social constructions. If the natural sciences have something to do with truth, it can only be truth in the sense of the truth as a social construction.

Against this background, the aim of the social constructivist analysis of the dynamics of physical research can best be understood: the objective is to show that physics are a product of the free choices of persons, a cultural product like any other. Or, as Pickering has formulated at the end of his defence of the social constructivist dynamics as motor of the development of the research in particle physics:

---

[6] SOKAL (1998) 913-914.

> «[…] there is no obligation upon anyone framing a view of the world to take account of what twentieth-century science has to say. […] World-views are cultural products; there is no need to be intimidated by them.»[7]

However, all this depends on whether the phenomena are more likely to be regarded as genuine discoveries, or rather as constructions. Are physical phenomena thus discovered or constructed? Is the worldview of physics a work of art that brings us no knowledge about a «real» order of nature? Or is it a worldview that we must interpret as a realistic description, and which certifies an independently existing rational order of nature?

These questions will be dealt with in the present study.

4. The Structure of this Investigation

To complete this introduction, let us now describe the structure of this investigation. The book is divided into five chapters: The first chapter aims to summarize the philosophical background of the debate between realism and social constructivism with respect to the dynamics of scientific research. The second chapter contains an introduction to astroparticle physics, together with some reflections on the aspects of this discipline, which may be relevant to our discussion. In the third and fourth chapter some of the issues indicated in the second chapter are examined in more detail. To this end, the work of two major research collaborations will be presented in detail: the MAGIC and the IceCube collaboration. In the fifth chapter the philosophical conclusions of this study are drawn.

A summary of these chapters is as follows:

(1) The Realism Debate

First, some episodes of the realism debate are summarized with the aim of highlighting the impact of this debate on the development of physics and of defining the part of the discussion that should be examined in this book: the dispute between realism and social constructivism. Immediately after that, the main positions in this part of the discussion are presented. Afterwards, we expose the consequences for our understanding of the reality in connexion with the topics that should be examined in this study, and also the previous history of the dispute is shortly summarized. Finally, some considerations are to be made, concerning the question how astroparticle physics can be used in order to win new impulses for our discussion.

---

[7] PICKERING (1984) 413-414.

(2) Astroparticle Physics: An Overview for Philosophers

First, some of the most important phenomena investigated by the astroparticle physics are presented. Immediately afterwards, various experiments and techniques are outlined which are now being used for studying these phenomena. Then follows a sketch of the theoretical models that are at present available to explain the phenomena of astroparticle physics. In addition, some results are mentioned that certain groups of theorists expect from the study of certain phenomena in this field. And finally, some preliminary notes for our analysis of the dispute between realism and social constructivism are derived.

(3) MAGIC

First, the structure of the MAGIC Collaboration is outlined. Immediately afterwards, the MAGIC Telescopes and some details of their technical development are presented. Then follows a description of the two variants of data analysis used by the members of the MAGIC Collaboration. Finally, further evidence for our analysis of the dispute between realism and social constructivism are collected on the basis of the issues presented in this chapter.

(4) IceCube

First, some of the social aspects are outlined which may potentially influence the work within a large scientific collaboration. Immediately afterwards, the IceCube Collaboration is presented. Then follows a description of the IceCube Detector as well as the procedure of data analysis which is used by members of the IceCube Collaboration. In addition, the predecessor of the IceCube Experiment – the AMANDA Experiment– is described briefly. Furthermore, some examples of the technical development of the components of the IceCube Facility are exposed, though not as detailed as the examples in the third chapter. Finally, further evidence for our analysis of the dispute between realism and social constructivism are collected on the basis of the issues presented in this chapter too.

(5) Conclusions and Outlook

First, a preliminary answer to the question of the suitability of the constructivist and the realist perspective to describe the dynamics of research in the field of astroparticle physics is given on the basis of the examples presented in this study. Then some examples of possible future developments in research are summarized which could confirm (or question) this answer. Following this, the conclusions drawn here are examined critically. Finally, the question is discussed, to what extent the results in the field of astroparticles are transferable to other fields of physics, and the consequences of this transfer for the discussion about the rational order of nature are specified.

It will be concluded that the dynamics of current research in astroparticle physics meets the description of the development of particle physics made by Brigitte Falkenburg:

> «The symbiosis of theory and experiment is the result and not the starting point of [scientific] development.»[8].

However, it will be also suggested that this situation could change in the future, if physicists were to lose confidence in the philosophical assumption of a natural order, independent of but accessible (at least in part) to human intelligence.

---

[8] Falkenburg (2007) 123.

1. Chapter: The Realism Debate

1. Introduction

Are the physical models and theories to be interpreted as realistic or anti-realistic (in an instrumentalist, fictionalist, positivist way, etc.)? That is, can we assume that physics provides (approximate) descriptions of the real, causal structures of nature? Or should we rather understand physical models and theories as useful mathematical fictions? They could be useful, because they help to report about certain recurring phenomena in a very condensed form, or because they constitute a fictitious view on the world, i.e., a habitable scenario where explanations can be found, which, however, do not have to correspond with reality.

These questions open up the so-called «realism debate»[9]. A discussion that has been accompanying physics over the centuries until today. This book is an attempt to get new impulses for the realism debate from the analysis of astroparticle physics. But before presenting some aspects of this research field it is necessary to deal with two tasks: first, it must be explained why it makes sense to continually deal with the realism question instead of excluding it as an old, intractable philosophical puzzle. And beyond that it is necessary to select some –as few and concrete as possible– of the topics discussed during the extended realism debate. Because numerous approaches have already been proposed and defended through the history of this controversy, and thus, an examination of all aspects of the debate in context of a book should necessarily be made in a very general way. Instead, it seems to me to be more revealing to focus on the analysis of a single current approach.

In the first chapter of this study, we will deal with the just-mentioned tasks. To this end, the chapter is subdivided as follows: first, [section 2] we summarize some episodes of the realism debate in order to emphasize both, the influence of this debate on the development of physics and the part of the discussion which we will examine in detail in this book. Immediately thereafter, [section 3] an account of the main positions in this partial discussion follows. After this presentation, it is already possible to discuss the serious consequences for our understanding of reality linked to the subjects to be examined here [section 4]. The following section [section 5] briefly summarizes the course of this discussion up to now. And finally, [section 6] there are some considerations about how new impulses for this debate can be won from astroparticle physics.

---

[9] See STÖCKLER (2003) for an overview of the current discussions about realism.

2. The Realism Debate in the History of Physics

It is not exaggerated to assert that the realism debate represent both an inseparable concomitance of mathematical physics and an important motor for its development. This claim can be confirmed through the examination of some key moments in the history of science.

Such a moment is undoubtedly the birth of mathematical physics. If we leave the initial explorations of the Pythagorean School aside, from which only a few fragmentary data were preserved we can put the birth of this discipline in the Platonic Academy (4th century BC). Because in the Academy it was –possibly for the first time– raised the problem of finding a simple mathematical model which is able to explain some apparently very irregular movements –namely the movements of the planets as seen from our perspective–.

This task resulted from the request of Plato (427-347 v. Chr.) to bring the movements of the planets into agreement with his ontology. Concerning this point it should only be mentioned that according to Plato the variable world of our experience –a world characterized by the constant generation, developing and annihilation of all things– should be understood as a copy of another, more fundamental kind of reality –the World of Ideas–. The World of Ideas is a timeless, structured, hierarchical reality with proportions and harmony, and hence entirety understandable. Unfortunately, the material universe only represents an inferior copy of this ideal reality, because the matter (from which the copy must be formed) entails temporality, and thus a constant movement and transformation of some entities into others. According to Plato this variableness makes it impossible to develop a science of material bodies, except in those cases where the objects somehow reflect a hint of the rational beauty (i.e., harmony, order, simplicity, etc.) of their ideal archetypes. In such cases, mathematics provides the language in which the connection between the material objects and the ideal archetypes can be expressed.

The celestial bodies however, are the material objects which are most similar to the entities of the World of Ideas, and consequently there should be a mathematical description of the motion of these bodies. But what form should this description have?

Since the World of Ideas is timeless and far away from all types of changes, the celestial bodies and their movements are to be described through simple and symmetrical forms that minimize the kinds of changes allowed in the movement. Because only those forms could characterize the things that are the least submitted to the yoke of matter. But the sphere and the circle are precisely the forms that meet these requirements best. Therefore, the following conclusions can be drawn from the Platonic view on the nature of the celestial bodies: (1) all celestial motions are circular, and (2) the rotational speed of the celestial bodies remains unchanged in these circular movements.

Now, the problem arises of how such theoretical considerations should be compatible with the reality of the complicated orbits. And so Plato formulated the question that should lead to the first model of mathematical physics:

> «What are the circular uniform and perfectly regular movements, which one should assume as hypotheses to save the phenomena of the planet?» [10]

Plato himself did not give a concrete answer to this question. But his pupil Eudoxus took on the task and worked out a cosmological model which showed that it is possible to describe such complex physical phenomena as the movements of the planets by simple formulas. In the words of Kanitscheider:

> «[Eudoxus] has now sketched for the first time a system of the planetary motions, which was suitable to reproduce the actually observed "crazy" retrograde and loop movements and the stop of the planets [...]. For each planet, the system consisted of a series of homocentric spherical shells around the Earth, where the outermost was formed by the fixed-star sphere; the inner spheres were superimposed having rotary axes with different directions, but all passing through the center of the Earth. By suitable variation of the direction of the axes and the rotation speed, it was possible for the planet itself, which was placed on the innermost shell above the Earth, to reproduce approximated by a Hippopede these loop-shaped movement that we actually observe in the wandering stars on the sphere» [11]

The cosmological model of Eudoxus, later further developed by Callippus, could to a certain degree describe the observed dynamics of the planets. But how should this model be interpreted? And, in particular, what to think about the invisible spherical shells of the model? Are they material, transparent bodies? Or are they mere mathematical tools without physical meaning? At this point, the opinions of Plato's disciples were divided. While Eudoxos did not attribute a physical meaning to the spheres, Aristotle thought it necessary to interpret these spheres as material bodies. Kanitscheider writes about this:

> «[...] the Eudoxus-Callippus model [was] certainly conceived by its creators as a mathematical tool for description [...],which made possible the interpretation of the apparent movement as the true movement appearing to the observer.
>
> [...] Eudoxos' and Kallippos' system had to be specially drawn up for each planet; the different spheres groups were not connected, and in addition stood the shells in no definable relation to reality. They were merely components of

---

[10] SIMPLICIUS, quoted in RIOJA - ORDOÑEZ (1999) 35-36.
[11] KANITSCHEIDER (2002) 62-63. (My translation). Interesting details of the cosmological model Eudoxos' can be found in TORRETTI (2007).

a mathematical construction, which had purely descriptive, but no explanatory function. This was for Aristotle (384-322 BC) an intolerable situation. It is thanks to him that the requirement was first established that cosmology, physics and philosophy cannot fall apart. A coherent picture of the world must reflect the real structures of the universe, based on true physical principles that are in turn consistent with the philosophical presuppositions.»[12]

Aristotle therefore modified the sphere model of Eudoxus by introducing additional spherical shells with the aim to interpret the spheres as real substances:

«First, he improved the cinematic properties in order to take better account of the phenomena; second, he materialized the model by regarding the spheres as bodies. [...]

While Eudoxus had never spoken about the cause of the planetary motions, Aristotle was concerned with the transmission of the rotational movement of the outermost sphere to the interne spheres. To this end, he had to suppose compensation spheres to avoid the interference of the proper motions of the planets [...].»[13]

It is noteworthy that the approaches of Eudoxus and Aristotle are evaluated differently by diverse present-day specialists. While authors such as Kanitscheider regard the contribution of Aristotle as a significant improvement of the original model of Eudoxus[14], other authors such as Torretti totally disagree[15]. The reason for this remarkable disagreement is that the schism that divided the Platonic Academy with regard to the interpretation of the model of Eudoxus still exists. What splits the different authors is their position in the dispute between realism and anti-realism, a dispute that clearly began with the birth of mathematical physics and still divides the minds of scholars.

Thus, from the very beginning the various theories and models of physics have always been accompanied by such a conflict of interpretation. Let us just

---

[12] Ibidem 64-65. (My translation). In my opinion one could object to this characterisation of the position Eudoxos' that for a Platonist the reality of the planetary spherical shells does not depend necessarily on their materiality. Because the mathematical objects are more real from a Platonic viewpoint as the material things. Therefore Eudoxos could deny that its construction stands «in no definable relationship with the reality». If this should be the position Eudoxos', then it would be an error to equate it with a mere instrumentalism. But such details are not very significant for the following argumentation. Because regardless of whether Eudoxus was anti-realist or realist with regard to his cosmological model, it is a fact that both options stood from the outset open.

[13] Ibidem 67. (My translation).

[14] Kanitscheider claims for example that the variant of Aristotle, «is the first real model of cosmology». Ibidem 67.

[15] See e.g. TORRETTI (2007).

consider another example: If one browses through the book which marks the beginning of the modern cosmology –*De Revolutionibus* by Copernicus– including the anonymous preface, added by Andreas Osiander to the text of the first edition (1543) of this work one finds, separated only by a few pages, these two radically different views about the meaning of the mathematical models of physics.

The anti-realistic (in this case, instrumentalist) opinion –represented here by Osiander's preface–, states that the Copernican Model is merely a calculation scheme that can describe the loops of the planets in the night sky properly, but still has no ontological significance. In his words:

> «For it is the duty of an astronomer to compose the history of the celestial motions through careful and expert study. Then he must conceive and devise the causes of these motions or hypotheses about them. Since he cannot in any way attain to the true causes, he will adopt whatever suppositions enable the motions to be computed correctly from the principles of geometry for the future as well as for the past. […] For these hypotheses need not be true or even probable. On the contrary, if they provide a calculus consistent with the observations, that alone is enough. […] So far as hypotheses are concerned, let no one expect anything certain from astronomy, which cannot furnish it, lest he accept as the truth ideas conceived for another purpose, and depart from this study a greater fool than when he entered it.»[16]

Copernicus himself had a very different opinion. He defended the realistic view that a good description of the phenomena can only be successful if the model characterizes the real structure of the world (more or less) correctly. Copernicus therefore argued that the problems of the Ptolemaic Astronomy arose from some incorrect assumptions about the structure of the world. And so the dedication of his book (to Pope Paul III) includes the following lines:

> «Nor could they elicit or deduce from the eccentrics the principal consideration, that is, the structure of the universe and the true symmetry of its parts. On the contrary, their experience was just like someone taking from various places hands, feet, a head, and other pieces, very well depicted, it may be, but not for the representation of a single person; since these fragments would not belong to one another at all, a monster rather than a man would be put together from them. Hence in the process of demonstration or "method", as it is called, those who employed eccentrics are found either to have omitted something essential or to have admitted something extraneous and wholly irrelevant. This would not have happened to them, had they followed sound principles. For if the hypotheses assumed by them were not false, everything

---

[16] Available at: http://www.webexhibits.org/calendars/year-text-Copernicus.html XIX.

which follows from their hypotheses would be confirmed beyond any doubt».[17]

The difference between the two interpretations of the Copernican Model is so radical that the rejection and even irritated reaction of the friends of Copernicus to the preface of Osiander is not surprising:

«The Copernicus circles reacted to Osiander's anonymous preface with anger and horror. [...] Even later partisans (such as Johannes Kepler) understood this as a treacherous falsification of the Copernican intentions.»[18]

Galileo later joined this group of Copernican realists:

«What unites Galileo and Kepler is an unambiguous realistic view of science. He wanted not only to save appearances, but to make true statements about physical bodies. Therefore, he also does not maintained the Copernican doctrine *ex suppositione*, following the example of Osiander and as Cardinal Bellarmine recommended it, but in full knowledge of the semantic function of factual hypotheses with a real referent, as a representation of reality. [...]»[19]

At this point it is important to emphasize two things:

(1) The adoption of a realistic rather than an anti-realist interpretation of the models of mathematical physics (or *vice versa*) cannot be understood as a matter of taste without further consequences. Instead, realism in the case of Copernicus (as in the case of Aristotle, Galileo and Kepler) was a strong motivation for the development of new physical ideas. At this point it should be merely noted that the realistic attitude Copernicus' concerning his cosmology forced him to face the problem of how the movement of the Earth could be physically possible. Copernicus himself has only carefully expressed a few conjectures concerning the physical issue[20], but even these meager indications showed very clearly that the heliocentric astronomy could be interpreted realistically only by overcoming Aristotelian physics. And this fact was one of the main motivations for Galileo's critical attitude regarding Aristotelian Physics[21].

(2) Both the realist and the anti-realist position have been represented through the history by a number of important physicists. This remark is appropriate here; because the realistic attitude of the founders of modern physics (Copernicus, Kepler, Galileo, Descartes, Newton ...) may give the impression that all physi-

---

[17] Available at: http://www.webexhibits.org/calendars/year-text-Copernicus.html XXI
[18] CARRIER (2001) 131. (My translation).
[19] KANITSCHEIDER (2002) 113. (My translation).
[20] See cap. 8 and 9 of the first book of *De revolutionibus*.
[21] See RIOJA - ORDOÑEZ (1999) 118-131 and cap. 3-4.

cists are realists and that the alternative positions enjoy popularity only among philosophers. In relation to this it must be said, however, that although the realism is a very widespread opinion among physicists –indeed, this approach could be called the «mainstream philosophy» of the physicists–, the alternative opinion also has been supported by specialists (such as Mach, Duhem, Gibbs, Deville, or to some extent, Feynman) that have made important contributions to the development of physics.

The previous notes can be supported with the help of a further historical example: the emergence of the modern atomic theory.

Nowadays hardly a physicist will doubt the real existence of atoms. But in the nineteenth century such a consensus did not exist. In Pierre Gassendi's and Robert Boyle's time (17th century) some authors already worked on the basis of the atomic hypothesis. And these efforts began to bear fruit at the beginning of the nineteenth century with Dalton's contributions to the development of chemistry. But although the progress of chemistry in the nineteenth century was often based on the assumption of the atomic hypothesis, Duhem showed that chemistry of his time could be formulated without this hypothesis at the end of the century [22].

Duhem himself supported an instrumentalist view of the physical (and more generally, scientific) theories:

> «Concerning the very nature of things, or the realities hidden under the phenomena we are studying, a theory conceived on the plan we have just drawn teaches us absolutely nothing, and does not claim to teach us anything. Of what use is it, then? What do physicists gain by replacing the laws which experimental method furnishes directly with a system of mathematical propositions representing those laws?
>
> First of all, instead of a great number of laws offering themselves as independent of one another, each having to be learnt and remembered on its own account, physical theory substitutes a very small number of propositions, viz., fundamental hypotheses.»[23]

That is, from this perspective one could interpret the atomic hypothesis (at best) as a useful fiction, with whose help a mathematical structure can be formulated, which can summarize and represent some experimental laws economically. In the writings of Ernst Mach we find the same view on the scientific theories as economic representations of the phenomena, supplying, however, no knowledge about the unobservable realities:

---

[22] DUHEM (2002).
[23] DUHEM (1991) .21.

> «According to Mach, physical theories only aim at an economical representation of experimental phenomena and the assumption of atoms does not belong to such an economical representation.
>
> In contrast, the founders of 20th century physics defended scientific realism. They were convinced that physical quantities and theories aim at the essential features of things and causal processes in nature. Boltzmann, Planck, Einstein, Rutherford and Bohr were atomist, personally participating decisively in the investigation of atoms.»[24]

Thus, at the beginning of the twentieth century a new episode of the dispute between realists and anti-realists takes place. And this episode enables us to confirm some of the aforementioned aspects of the discussion:

(1) First, we see that while the mainstream of the key participants assume a realistic attitude about the existence of atoms, the weight of the representatives of the anti-realism is also considerable. This shows that the realism debate is not a contrived philosophical diversion but corresponds to a tangible schism in the scientific community.

(2) And secondly, we can state that the adoption of a realistic or an anti-realist attitude undoubtedly affects the further development of physical theories. Because representatives of atomism contributed decisively in this period to the development of some areas of physics such as the kinetic theory, electromagnetic theory, and spectroscopy, while many anti-realists preferred the thermodynamics as a «metaphysics-free» form of physics:

> «There is no doubt that those wishing to make a case for atoms were able to steadily strengthen their case during the closing decades of the twentieth century. However, it is important to put this in perspective by taking account of spectacular developments in thermodynamics which were achieved independently of atomism, and which could be, and were, used to question atomism, branding it as unacceptably hypothetical.
>
> Phenomenological thermodynamics, based on the law of conservation of energy and the law ruling out spontaneous decreases in entropy, supported an experimental program that could be pursued independently of any assumptions about a micro-structure of matter underlying properties that were experimentally measurable. The program was developed with impressive success in the second half of the nineteenth century.»[25]

---

[24] FALKENBURG (2007) 46.
[25] CHALMERS (2005).

An increasing number of indications –such as for example the experiments with cathode rays by Thomson and Millikan or the confirmation of the atomistic description of the Brownian motion by the experiments of Perrin– persuaded the very most physicists in the first decades of the twentieth century that there really are atoms. But because the existence of atoms can be derived only indirectly from the observable phenomena, the position of the physicists appears only meaningful in the context of a realistic attitude regarding the unobservable entities included in the successful models of physics. Therefore numerous authors judge the acceptance of the modern atomic theory as a triumph of scientific realism.

Nevertheless the ideas of Duhem and Mach about the knowledge potential and goals of the physical theories were not regarded simply as rejected. Because the anti-realistic –in this case instrumentalist and empiricist– attitude of these authors crucially influenced the important epistemological and science-theoretical school of the «Vienna Circle» which is considered the germ cell of logical empiricism. This positivist school emphasized the role and the special epistemological status of the observable phenomena by undertaking the task of reconstructing the meaning of all meaningful concepts of science in terms of pure observation statements. This way, they tried to draw a clear boundary between physics and metaphysics in order to achieve a flawless phenomenal knowledge without metaphysical contamination (and in particular without the assumption of the existence of the unobservable entities).

So, do physics –and the other natural sciences– only serve to acquire knowledge about observable phenomena or does it also provides approximately true descriptions of the unobservable entities postulated in the successful models of this phenomena? Much of the debate between realists and anti-realists of the 20th century has been oriented to the question of whether it is really possible in physics to draw an exact epistemological line between empirical facts and theoretical (even «metaphysical») elements of the models. Is it possible in the context of physics to find «metaphysical-free» phenomena at all? And even if this should be the case, is it meaningful to attribute a completely different epistemological status to different (empirical and not empirical) aspects of the same physical model?

For strict empiricists like van Fraasen only those substructures of the physical models are epistemically relevant, which are isomorphic to the observable phenomena. That is, «science aims to give us theories which are empirically adequate; and acceptance of a theory involves as belief only that it is empirically adequate»[26]. On the other hand, realists like Sellars support the view that «to

---

[26] VAN FRAASEN (1980) 12.

have a good reason for holding a theory is ipso facto to have a good reason for holding that the entities postulated by the theory exist»[27].

In considering such a state of affairs, one might be inclined perhaps to draw the conclusion that in the physical and philosophical discussion of the last century concerning scientific realism no decisive intellectual novelty has emerged. For the just-mentioned contrasting positions are very close to their classical counterparts. (Compare, for the example, the just cited quotation from van Fraasen with the above-described approach of Osiander). But such verdict would be premature, because in recent decades a completely new variant of the anti-realism has emerged: the social constructivist interpretation of dynamics of scientific research.

The next section is dedicated to the presentation of this variant of anti-realism, and we will again meet the social constructivist interpretation of history of physics in many places of the present study. However, it is appropriate to make two remarks about this new anti-realistic trend already at the end of this section:

(1) It must be noted first that the debate between realism and social constructivism –as the other variants of the realism debate– is not a purely philosophical discussion of little interest for physicists. In fact, the most important social constructivist study of dynamics of scientific research is a book about the history of the standard model of particle physics that has been written by a physicist who contributed to the development of this model. And the most important critics of this author are also physicists.

(2) It must also be noted that the social constructivist challenge is far more radical than the traditional criticism of realism on the part of empiricists and instrumentalists. As far-reaching the difference between the realistic and the instrumentalist view on mathematical models of physics may be, both views share the opinion that physics provide us with some rational knowledge about nature; knowledge that is obtained through a combination of technical arguments and experimental tests. But both, technical arguments and empirical evidence (and even nature itself!) play a very minor role in constructivism. In Franklin's words:

> «Social constructivists imply, however so much they may disclaim it, that science does not provide us with knowledge. [...] In constructivist case studies of science, experimental evidence never seems to play any significant role. In their view the acceptance of scientific hypotheses, the resolution of discordant results, as well as the acceptance of experimental results in general, is based on "negotiation" within the scientific community, which does not include evidence or epistemological or methodological criteria. These negotiations do include considerations such as career interests, professional

---

[27] SELLARS (1962) 97.

commitments, prestige of the scientists' institutions and the perceived utility for future research.»[28]

That approach is so new, and it can undermine the cognitive pretensions of physics so crucially, that it is worthwhile to discuss their main features in more detail. This is the goal of the next section.

3. The Social Constructivist View on the Development of Physics

It is often problematic to determine the exact starting point of a school of thought or a research program in the history of philosophy. But in this case one can justifiably regard the social constructivist interpretation of the development of physics as a result (unintended by its author) of the analysis of scientific revolutions in *The Structure of Scientific Revolutions* (1962) by Thomas Kuhn. Alexander Bird explained this as follows:

> «A central claim of Kuhn's work is that scientists do not make their judgments as the result of consciously or unconsciously following rules. Their judgments are nonetheless tightly constrained during normal science by the example of the guiding paradigm. During a revolution they are released from these constraints (though not completely). Consequently there is a gap left for other factors to explain scientific judgments. Kuhn himself suggests in *The Structure of Scientific Revolutions* that Sun worship may have made Kepler a Copernican and that in other cases, facts about an individual's life history, personality or even nationality and reputation may play a role [...]. Later Kuhn repeated the point, with the additional examples of German Romanticism, which disposed certain scientists to recognize and accept energy conservation, and British social thought which enabled acceptance of Darwinism [...]. Such suggestions were taken up as providing an opportunity for a new kind of study of science, showing how social and political factors external to science influence the outcome of scientific debates. In what has become known as social constructivism/constructionism [...] this influence is taken to be central, not marginal and to extend to the very content of accepted theories»[29].

Although Kuhn distanced himself explicitly from the constructivist positions it is undisputed that after the publication of his book a number of social constructivist studies have appeared, which were understood by their authors as unfolding (or confirming) Kuhn's ideas[30].

---

[28] FRANKLIN (1999) 4.
[29] BIRD (2004).
[30] See e.g. PICKERING (1984) 407-411.

The history of social constructivist approaches in the philosophy of physics is therefore relatively short; it goes back to the seventies of the last century. However, in this period there have been numerous interesting contributions, which have set different accents in the constructivist analysis. As it is not unusual in the development of a school one can say that there are almost as many variants of the social constructivist program as authors who have participated in its development.

For this reason –and in order to ensure that the approach of this school, used as initial position for the coming discussion, is clearly formulated– I focus here on a particular version of the constructivist discourse: the version of Andrew Pickering. The reason for this decision is clear: the studies of Pickering (especially his book *Constructing Quarks*) together with the studies of Collins, to which we will also pay particular attention in the following chapters, are the most detailed social constructivist analysis of dynamics of physical research. And they are generally recognized as an important milestone of this program.

So we try to synthesize the picture of the development of physics suggested by Pickering and his colleagues. The main thesis of this approach is clearly anti-realistic and can be summarized as follows:

> –The physical theories are merely cultural constructions that arise from the negotiations between various interest groups, and do not describe such thing as a «real structure» of the world. They do not have thus to be taken into account in the development of a world view.

For Pickering this thesis is relevant in such a way that it is mentioned several times during development and at the end of his book *Constructing Quarks*. As a kind of conclusion of his study Pickering devotes the last thoughts of this work to underline the mentioned thesis once again by contrasting it with the realistic attitude of John Polkinghorne:

> «Many people do expect more of science than the production of a world congenial to social understanding and future practice. Consider the following quotation: "Twentieth-century science has a grand and impressive story to tell. Anyone framing a view of the world has to take account of what it has to say... It is a non-trivial fact about the world that we can understand it and that mathematics provides the perfect language for physical science: that, in a word, science is possible at all."[31]
>
> Such assertions about science are commonplace in our culture. In many circles they are taken to be incontestable. But the history of HEP [high energy physics] suggests that they are mistaken. [...] On the view advocated [here],

---

[31] POLKINGHORNE (1983) 24.

there is no obligation upon anyone framing a view of the world to take account of what twentieth-century science has to say. [...] In certain contexts, such as foundational studies in the philosophy of science, it may be profitable to pay close attention to contemporary scientific beliefs. In other contexts, to listen too closely to scientists may be simply to stifle the imagination. Worldviews are cultural products; there is no need to be intimidated by them.»[32]

Certainly no realist would deny that scientific theories are cultural products. Because they are not announced by angels but the result of the efforts of scientists with different professional and personal backgrounds and interests. But from a realistic point of view such a social component of the research processes does not prevent that the results of these processes represent (approximate) descriptions of the real structure of the world −descriptions that must be thus taken seriously−. A realist would therefore challenge the idea that a cultural product can not claim to be true. But Pickering and many other constructivists argue that way. Why?

The reason for this assertion lies in the constructivist analysis of the relations between experiments and theories of physics. The main line of the argument can be summarized as follows:

1. The experiments cannot just confirm or reject theories (or models) because the experimental results must always be interpreted and evaluated.
2. Each interpretation and each evaluation contains judgments which are not based solely on facts. The scientists thus make active decisions which substantially determine the development of the sciences.
3. These decisions are best understood from the sociological perspective. (That is, they are the results of interactions between certain social groups −diverse groups of theorists and experimentalists, etc.−, which pursue specific interests and goals).
4. Consequently, the physical theories are cultural constructions which arise from the interaction between different groups of interests and not from a process of mapping the actual structures of the world.

Some comments from Pickering concerning items 1-3 are as follows:
On 1. and 2.:

«There are, though, two well-known and forceful philosophical objections [to the view that experiment is the supreme arbiter of theory], each of which implies that experiment cannot *oblige* scientists to make a particular choice of theories. First, even if one were to accept that experiment produces unequivocal fact, it would remain the case that choice of a theory is underde-

---

[32] PICKERING (1984) 413-414.

termined by any finite set of data. It is always possible to invent an unlimited set of theories, each one capable of explaining a given set of facts. Of course, many of these theories may seem implausible, but to speak of plausibility is to point to a role for scientific *judgment*: the relative plausibility of competing theories cannot be seen as residing in data which are equally well explained by all of them [...]

The second objection [...] is that the idea that experiment produces unequivocal fact is deeply problematic. [...] [If] everyone agrees upon how an experiment works and that it has been competently performed, there is no way in which its findings can be disputed. However, it appears that this is not an adequate image of actual experiments. They are better regarded as being performed upon "open", imperfectly understood systems, and therefore experimental reports are *fallible*. This fallibility arises in two ways. First, scientists' understanding of any experiment is dependent upon theories of how the apparatus performs [...]. More far reaching than this, though, is the observation that experimental reports necessarily rest upon incomplete foundations. [...] [Much] of the effort which goes into the performance and interpretation of HEP experiments is devoted to minimizing "background" [...]. Experimenters do their best, of course, to eliminate all possible sources of background, but it is a commonplace of experimental science that this process has to stop somewhere if results are ever to be presented. Again a *judgment* is required, that *enough* has been done by the experimenters to make it probable that background effects cannot explain the reported signal, and such judgments can always, in principle, be called into question».[33]

On 3.:

«The scientist legitimates scientific judgments by reference to the state of nature; I attempt to understand them by reference to the cultural context in which they are made. I put scientific practice [...] at the center of my account. [...] My goal is to interpret the historical development of particle physics, including the pattern of scientific judgments entailed in it, in terms of the dynamics of research practice».[34]

«My analysis of the social coherence of scientific judgments has focused upon research traditions and their symbiosis. My suggestion has been that the history of HEP should be seen *in terms of* a fluctuating pattern of *research*

---

[33] PICKERING, A. (1984) 5-6.
[34] Ibidem 8.

traditions, wherein coherent sets of judgments were structured by *shared* sets of research resources.».[35]

Pickering puts here a crossroad between his view of the development of physics and the realistic approach, he calls «scientific account». Therefore, it is necessary to include a brief summary of dynamics of experiment and theory and the related interaction between scientific groups (or «traditions») on Pickering's account.

According to this account one finds the key to the «scientific judgments», necessary for the interpretation of experiments in physics on the interaction between the so-called «experimentalist traditions» and the «theoretical traditions». The experimentalist traditions form around different observation techniques and instruments that exist at a particular point in time. The theoretical traditions are based on various mathematical techniques and theoretical previous work, also available in a particular context. Based on the existence of such traditions, Pickering states first that the teams of theorists and experimentalists usually have a symbiotic relationship:

> «Consider the theoretical tradition: to justify his choice to work within it, a theorist has only to cite the existence of a body of experimental data in need of explanation. And fresh data, from succeeding generations of experiment, constitute the subject matter for further elaborations of theory. Conversely, for the experimenter, his decision to investigate the phenomenon in question rather than some other process is justified by its theoretical interest, as manifested by the existence of the theoretical tradition».[36]

Up to this point, I think that a realist would hardly contradict Pickering's presentation. Because the existence of common interests of theorists and experimentalists is incontestable, and beyond that, it seems to stand in no contrast to a realistic understanding of the contents of the theories developed by this dynamics. But Pickering goes a step further, claiming that decisions in favor of particular interpretations of results of various physical experiments are based on the opportunities for a further development of their activities that experimentalists as well as theoreticians hope for.

A similar but somewhat weakened version of this statement –i.e. the thesis that the consideration of the chances for the further development of a research line can affect the interpretation of the results of some (or many) experiments– might still be acceptable for a realist. For scientists are not pure spirits, judging entirely without regard of advantages and disadvantages; moreover, one could also evaluate these new opportunities for a line of research as an indication for the truth of the favored interpretation. But Pickering's thesis is much stronger. It

---

[35] Ibidem 406.
[36] Ibidem 10.

does not see considering the chances for the future development of the research traditions as *a* factor (among others) but as *the* mechanism which determines the evolution of physics:

> «Whilst it is possible to argue that all experiments are in principle fallible, experimenters do not enter this as an explicit caveat when reporting their findings: they simply report that, say, a particular quantity has been measured to have a particular value (often within a stated margin of uncertainty). It is then up to their colleagues to decide whether or not to challenge them, *and such decisions are related to decisions over future practice*. If the theorist can find the resources for a constructive analysis of the data, one should not expect him to spend long periods of time in searching for ways to challenge the experiment. If the data, through the medium of theory, raise new problems for experiment to investigate, then neither will experimenters be disposed to probe deeply. In this way, the potential fallibility of experiments is rendered *manageable*. A parallel solution is available to the problem of the underdetermination of theory. While a multiplicity of different theoretical explanations may be conceivable for any set of data, not all of these will be attractive in terms of the dynamics of practice».[37]

At this point, a realist must already protest. For this description of the interaction between experimentalists and theoreticians has at least two deficits (from a realistic perspective):

(1) The weighing of advantages and disadvantages is considered as the last resort instead to acknowledge that the leading role of research dynamics lies in precise data collection procedures and detailed analysis of the data. This way there is the possibility that faulty conducted experiments are found to be correct when their results correspond to expectations and offer sufficient advantages. And this ought not to be seen as a regrettable anomaly but as one option among others about doing research.

(2) In describing the interactions between experimentalists and theoreticians the role of theoretical proposals is so stressed that an at least semi-autonomous development of the experimental research with regard to the theories under examination hardly seems to be conceivable. But from a realistic perspective it is important to ensure the relative autonomy of experimental research. As the experiments should in fact serve to clarify the characteristics of nature and provide the basis for the work of theorists as well as the tests of their hypotheses. Therefore, a realistic analysis of dynamics of physical research must assume that even if certain experimental results support a theoretical hypothesis opening up a new

---

[37] Ibidem 13.

interesting research field for experimentalists, the validity of these results will be further checked in every detail. And a realistic analysis of dynamics of physical research must additionally regard it as possible that such checks finally lead to some appreciated theoretical proposals being rejected, which were seen as experimentally supported at first. The rejection of such proposals will be based on reasonable (and from the theories under examination largely independent) investigations of experimental physicists.

At this point, one can clearly identify the difference between realistic and constructivist analyses of dynamics of research. According to the constructivist analysis both, experiments and experimental arrangements as well as related data analyses have a large plasticity. A plasticity that enables the «negotiations» between experimentalists and theorists. Of course, this plasticity is not absolute. Not every idea or every proposal can be confirmed immediately. But if a theoretical scenario is very attractive both, for theoreticians and experimentalists, one will probably find a way to discover (or more clearly said to «construct» or to «produce»)[38] experimental evidence for this or a similar scenario. This goal can be reached with a new evaluation of the known data (with the help of new methods of data analysis), a new theoretical model of the relevant measuring instruments or a modification of the experimental arrangements. On the way to this goal one will surely find «resistances» of the material world, which the physicists must overcome by «accommodations»:

> «The dance of agency, seen asymmetrically from the human end, thus takes the form of a *dialectic of resistance and accommodations,* where resistance denotes the failure to achieve an intended capture of agency in practice, and accommodation an active human strategy of response to resistance, which can include revisions to goals and intentions as well as to the material form of the machine in question and to the human frame of gestures and social relations that surround it».[39]

But at the end of this process there are some models of the relevant measuring instruments and some models of the examined phenomena as well as a number of different experiments and experimental procedures supporting each other and mainly corresponding to the interests of experimentalists and theorists. Also nature (unfortunately) interferes by means of the «resistances» in this process:

---

[38] «It is *unproblematic* that scientists produce accounts of the world that they find comprehensible: given their cultural resources, only singular incompetence could have prevented members of the HEP community producing an understandable version of reality at any point in their history». PICKERING (1984) 413.
[39] PICKERING (1995) 22.

> «In this way, then, *how the material world is* leaks into and infects our representations of it in a nontrivial and consequential fashion».[40]

But this interference is not crucial. In Franklin's words:

> «Although Pickering acknowledges the importance of the natural world, his use of the term "infects" seems to indicate that he isn't entirely happy with this. Nor does the natural world seem to have much efficacy. It never seems to be decisive in any of Pickering's case studies. Recall that he argued that physicists accepted the existence of weak neutral currents because "they could ply their trade more profitably in a world in which the neutral current was real." In his account, Morpurgo's observation of continuous charge is important only because it disagrees with his theoretical models of the phenomenon. The fact that it disagreed with numerous previous observations of integral charge doesn't seem to matter.
>
> [...] The natural world seems to have disappeared from Pickering's account.».[41]

From a constructivist point of view it is only reasonable that nature plays a subordinated role because in this context physics is not understood as a description of the world, but as a cultural invention, like arts. And this implies that, as soon as the artistic-theoretical proposals –clearly to be understood as constructions–, are accepted by a large number of specialists, they should dominate the whole dynamic of research, while the work of experimental physicists mainly is to supply a confirmation for these popular constructions by avoiding the natural «resistances». For that reason Pickering strongly criticized Gieres' view that the experimental resistances can let a theory appear far more attractive than others. In Pickering's words: «[...] here we are back in a good old-fashioned context of justification in which the data tell scientists what to believe»[42].

According to Pickering's view, the theories cannot in any case be so dependent on the resistances of nature.

Quite the opposite, a realistic analysis of dynamics of research entails the thesis that the experiments can be largely carried out and evaluated independently from the wishes and trends of theorists; the thesis that the crucial role in dynamics of research corresponds to the phenomena discovered and described in the experiments, the theories must adapt to; and the thesis that the process of testing the theoretical proposals in experiments is mainly influenced by rational arguments and not by calculation of possible advantages and disadvantages for future practice.

---

[40] Ibidem 183.
[41] FRANKLIN (2009).
[42] PICKERING (1991) 578.

We can better explain the differences between the realist and the constructivist analysis of dynamics of research with the help of a schema:

Structure of the relationship between theory and data

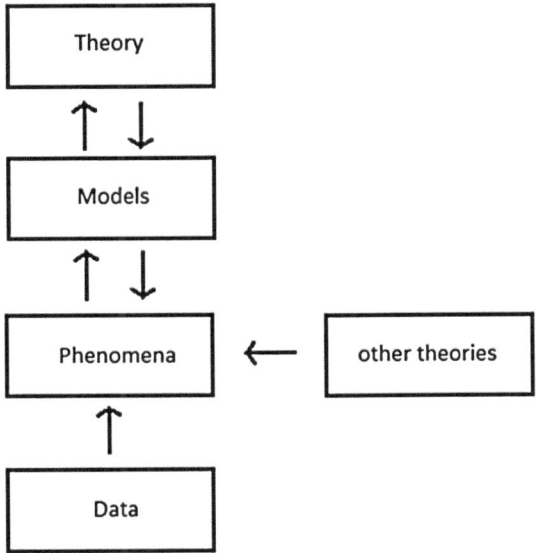

In this scheme, the main elements of the dynamics of theory and experiment are shown. At the top of the scheme are the theories in the testing phase. They are first to be understood as pure mathematical structures from which some parts can be related to the physical world in any way. But one usually connects the theories only indirectly with certain aspects of the physical world, that is, by mediating various models. The models themselves can also be understood as mathematical structures which include some additional assumptions in addition to the structural properties of the theories. Mathematically one can see the models as concrete instances of the theories. But the interesting thing about these structures is that they can at least partly be used for representing the characteristics of some physical systems, so that they are appropriate to explain physical phenomena. Theories and models thus represent the purely theoretical parts of the physical building. They are depending on one another: with the help of the fundamental theories, models for different entities and processes are built; but by using approximate models, the outlines of a requested but not yet formulated theory can be defined. It is especially important to emphasize, however, that models and theories are to be understood first as pure creations of the human mind or, if you will, as mere cultural products.

The lower segment of the scheme is focused on the concept of «phenomenon». The phenomena are the simple correlations between events, processes, etc., which can be found by analyzing the data. The phenomena should thus not be confused with the data (also called «raw data», «basic data» or «primary data»). Data usually stands for the numbers that produce the different detectors and measuring devices in each run[43]. Usually one gets different numbers in each individual run. But if one examines these fluctuating data with the help of an established theory regarding the measuring instrument plus a number of statistic procedures and other techniques of data analysis, some regularities may be discovered, which will later have to be explained with the help of the theoretical models and theories. These regularities which are discovered on the basis of the empirical data and the theoretical background knowledge are the phenomena.

On the basis of this schematic representation of dynamics of physical research it can be maintained that the main difference between social constructivism and scientific realism at this point is that from a constructivist perspective the represented dynamics are «top-down» controlled, whereas from a realistic view the phenomena are predominantly «bottom-up» discovered. That is, from the constructivist perspective the theoretical constructions –the products of creative minds– determine the development of physics, while the role of the experimentalists is confined to that of developing procedures for data collection and data analysis from which phenomena are derived, which should correspond to theoretical expectations as far as possible.

From the realistic perspective on the other hand, the phenomena are not constructed, but discovered with the help of already established measurement and analysis methods that were developed and tested independently from the theories and models under examination. Thus, experimental physics has large autonomy in relation to new theoretical impulses and finally ensures that only those theoretical proposals are accepted (or able to subsist) that can explain the physical properties of the real world.

For the purposes of our study it is neither necessary to elucidate the differences between the constructivist and realistic view of the dynamics of research nor to explain the principles of the social constructivist approach any further. In the final sections of the chapter I first summarize briefly the attempts that have been made to resolve the dispute between realism and social constructivism so far. Then I point to the impulses that (in my opinion) can be gained from astrophysics in order to bring this discussion a step forward. But

---

[43] But in some experiments, the data can be images, colors, photographic traces and so on.

first we should temporarily stop to illustrate the importance of the debate between realism and constructivism. That is the goal of the next section.

4. The Meaning of the Dispute between Realism and Constructivism

Is there any point in undertaking a long investigation in order to shed some light on the controversy between realism and social constructivism? Would it not be better to place, for example, other more common variants of anti-realism (such as the new versions of the empiricist approach) in the center of this study?

To show that it is really worthwhile to carefully examine the social constructivist analysis of dynamics of physical research, it is essential to emphasize the revolutionary radicalism of the social constructivist view as well as the important role this radical perspective can play –and to some extent already does– in the future development of our culture.

At the end of the second section of this chapter I already pointed out that the social constructivist challenge to realism is much more radical than the traditional criticism of the empiricists. The scope of this challenge is easier to understand when looking at the Galilean Program which essentially represents the mainstream[44] program of modern physics. The basic idea of the Galilean View can be found in a very famous (and in philosophical contributions of physicists often quoted) passage of *Il Sagiattore*:

> «Philosophy is written in that great book which ever lies before our eyes –I mean the universe– but we cannot understand it if we do not first learn the language and grasp the symbols, in which it is written. This book is written in the mathematical language, and the symbols are triangles, circles and other geometrical figures, without whose help it is impossible to comprehend a single word of it; without which one wanders in vain through a dark labyrinth»[45]

The representation of the whole of nature as a book written in the language of mathematics, physicists must learn to understand, was an audacious thesis at Galileo's time, because there was no mathematical description of most phenomena then. But the successful history of modern physics can be interpreted as an increasing confirmation of the Galilean View. In the 20th century Dirac, for example, wrote:

---

[44] One could almost say that this is the program of the *entire* modern physics. But with this statement we would forget the (above mentioned) contributions of the empiricist school of Mach, Duhem etc.
[45] Quoted in BURTT (2003) 75

> «It seems to be one of the fundamental features of nature that fundamental physical laws are described in terms of mathematical theory of great beauty and power, needing quite a high standard of mathematics for one to understand it. You may wonder: Why is nature constructed along these lines? One can only answer that our present knowledge seems to show that it is so constructed. We simply have to accept it. One could perhaps describe the situation by saying that God is a mathematician of a very high order, and He used very advanced mathematics in constructing the universe. Our feeble attempts at mathematics enable us to understand a bit of the universe, and as we proceed to develop higher and higher mathematics we can hope to understand the universe better»[46].

Two theses relevant for our discussion can be inferred from this and similar texts of prominent physicists: the Galilean Idea that the world is constructed mathematically and the realistic assumption that physical theories supply approximate (but always better) descriptions of this rational structure of the world. Most important of all is to emphasize what Dirac and most realists among physicists have in common: a realistic attitude in relation to the entities that are assumed and characterized in the models of physics requires the assumption that the world itself possesses a certain mathematical form. To give another example of this position Max Planck is quoted as follows:

> «In any case we can say in conclusion that, after all what teaches the exact natural sciences, over the whole realm of nature [...] a certain lawfulness reign, which is independent of the existence of [...] a thinking mankind. It therefore represents a rational world order, to which nature and humanity are subject, but whose essential nature is and remains unknowable to us [...]. But we are entitled by the actually rich achievements of scientific research to the conclusion that we are at least approaching constantly through relentless continuation of the work to the unattainable goals, and they reinforce our hope of a steadily progressive deepening of our insights into the reigning of the all-powerful reason which govern over nature.»[47]

In my opinion the mentioned passages of Galilei, Dirac and Planck very well summarize what might be called the «classical realist world view on physics»[48]:

---

[46] DIRAC (1963)
[47] PLANCK, M. (2001), *Vorträge und Erinnerungen* (Berlin, Springer 2001) 169. (My translation).
[48] Which does not mean that all physicists share this picture. As we have already mentioned in the 2nd Section of this chapter, there are also a group of eminent physicists who have adopted an anti-realist perspective. It is fair to say, however, that the majority of the physicists are closer to the «classical realist worldview of physics» than to the anti-realist positions.

the picture of a mathematically structured physical reality, whose structure becomes understandable (even if only approximately) step by step through the work of physicists. Beyond that, these passages suggest certain proximity of the physical worldview to natural theology. But exactly at that point it becomes obvious how strange such a picture must be for a growing number of people from cultivated classes of our present western civilization. This growing estrangement of the Galilean Views from the perspective of the current *Zeitgeist* did not do unnoticed by philosophers of physics, as Torretti confirms:

> «The founders of modern physics took Plato's metaphor [of the laws of nature] quite literally, and set out to find the articles of nature's legal code. In the writings of these Christian authors the word "law" does not signify the universal scope of the prescribed regularities, but rather the legislative authority of their divine source.
>
> [...] Today few would countenance basing physics on our knowledge of God. However, one implication of seventeenth-century theological commitments has lingered on as a source of confusion. Like the smile of the Cheshire cat, the idea of a ready-made world continues to haunt a philosophical tradition from which the idea of it Maker has long vanished. [...] the dream persists of a final theory of everything representing the true mathematical structure of the universe.»[49]

Well, I don't actually think that the idea of a Maker of the World has vanished from physical tradition. But regardless of how accurate this claim may be[50], Torretti's text clearly refers to an assumption of the physical worldview which encounters larger resistance in the philosophy nowadays, i.e. to the assumption that there is such a thing as «the true mathematical structure of the universe» which our theories must adapt to.

On the part of physics a number of authors tried to weaken this thesis (to some extent) in the last decades To this end, the idea has been developed that there may be more than one single mathematical structure that can be accepted as an appropriate characterization of physical reality. To use an anatomical metaphor, according to these authors it might be justified, to divide the nature in different ways, as long as the cuts are made at natural joints. Or in other words:

---

[49] TORRETTI, R. (1999), *The philosophy of physics* (Cambridge, Cambridge University Press) 406; 432-433.
[50] Already the mere fact of the continuous references to God in popular books of the specialists who try to explain the current physical (and specially cosmological) theories makes this claim of Torretti dubious. It seems that even those physicists who declare themselves «non-believers» do not stop imagining a kind of virtual God: The Mind responsible for the enigma that they try to solve.

one could argue that nature determines a class of possible theories but not a concrete structure within this class. As Franklin states:

> «It is doubtful that the world, or more properly, what we can learn about it, entails a unique theory. [...] [The rationalists] argue that the way the world is restricts the kinds of theories that will fit the phenomena, the kinds of apparatus we can build, and the results we can obtain with such apparatuses.»[51]

In my opinion most physicists could acquire a taste for such a plurality of possible structures of the world subject to certain conditions. But the very idea that there are such things as «natural joints» –that is, certain features, distinctions, constants, and facts that must be taken into account by the various physical scenarios because they represent real aspects of nature– is taken as an unacceptable limitation of creative freedom, or as a sign of patriarchal thinking, or even as a theological provocation by many philosophers and humanists of our time. Why? Because regardless of whether the world can be characterized by a single structure or a whole class of mathematical structures, it manifests itself as a reality pervaded by mathematical rationality which largely determined the «true» (and perhaps also the «beautiful») in an objective sense. In this scenario truth (and beauty) is in parts open to human reason but appears as something «given» which one must adapt to. John Polkinghorne who researched for many years as a coworker of P. Dirac explains this as follows:

> «Dirac made his many great discoveries in the course of a lifelong and highly successful quest for mathematical beauty.
>
> When we use abstract mathematics in this way as a guide to physical discovery, something very odd is happening. After all, mathematics is pure thought and what could it be that links that thought to the structure of the physical world around us? Dirac's brother-in-law, Eugene Wigner, who also won a Nobel Prize for Physics, once called this "the unreasonable effectiveness of mathematics." He also said it was a gift that we neither deserved nor understood.»[52]

Such an understanding of physics research and the connection between the world of thought [mathematics] and nature is closely related to venerable philosophical traditions, such as the Pythagorean-Platonic Tradition or natural theology. It is, however, in glaring contrast to some very popular philosophical opinions of our time, such as the assumption that rationality is something exclusively human and cannot be applied to nature, or the opinion that the mathematical theories are conventional constructions (such as games) which have no relation to the physi-

---

[51] FRANKLIN (2009).
[52] POLKINGHORNE (2003).

cal world, or the view that the cultural traditions (including science) form a landscape of ideas and approaches that complement each other, overlap, conflict, enter into dialogue with each other or ignore each other without any extra-mental reality (as a supposed «real structure of the world») that could lead to a gradual convergence of the different trends in this landscape. A view that goes back at least to the philosophy of Nietzsche, and which has been assumed by most of the theorists of the so called «postmodern thought».

In other words, the realist view of physics as a science that brings us gradually closer to the real structure of the physical world[53] can hardly be combined with the current dominant conception of the autonomy of the human (cultural) sphere in relation to the non-human areas of reality, a cultural sphere which is in no way related to a «natural order».

For this reason, various routes have been explored, in order to weaken the «metaphysical» aspects of the physical (and/or Galilean) program –in particular the idea of a mathematical order of the world– in the last century. But the recent contributions to this approach remain somewhat questionable. It was e.g. suggested that the rational order of nature developed (or could develop) from an original chaotic (that is lawless) state. But this idea is problematic[54]. All empiricist attempts to cleanly separate the empirical contents from the metaphysical requirements of physics have been just as problematic so far[55]. And other, for example, Neo-Kantian[56] attempts to provide an anti-realist or «anti-metaphysical» interpretation of the physical theories also had no enduring influence on the way physicists interpret their own work.

Under these circumstances, it is not surprising that physics (and all natural sciences) are increasingly being criticized by philosophers and other humanists for whom the existence of an objective rational order –and even a kind of «legality»– in nature almost is intolerable. When taking this into account the attitude of the humanists is understandable. Let us remember and complete the quotation of Franklin at the introduction of this book:

---

[53] Or to the class of structures that can be regarded as a representation of the material world.

[54] For this proposal see e.g. HELLER (2003) chap.11. The main problem for the consideration of the laws of physics as product of an irrational chaotic state is that even the physical chaos is not equivalent to a lawless «philosophical» chaos. A physical chaos is always subject to some regularity. It requires at least the validity of the probability theory.

[55] After a decades-long discussion, the so-called logical empiricism is today generally seen as failed. At this point, see e.g. BROWN (1993) part I. But also the latest attempt of an empiricist reading of the physics (the «constructive empiricism» of van Fraasen) is very problematic. See e.g., to this point FRANKLIN (1999) 149-162 and also FALKENBURG (2007) chap. 1.

[56] The main neo-Kantian alternative to realism today is the so-called «internal realism» of Putnam. A criticism of this approach can be found for example in DEVITT (1991) 113-137.

«The attitude of many humanists toward science has changed from indifference to distrust, and even hostility. Today science is under attack from many directions, and each of them denies that science provides us with knowledge. Whether it is because of inherent gender bias, Eurocentrism, or the social and career interests of scientists, science is untrustworthy and fatally flawed. "The radical feminist position holds that the epistemologies, metaphysics, ethics, and politics of the dominant forms of science are androcentric and mutually supportive; that despite the deeply ingrained Western cultural belief in science's intrinsic progressiveness, science today serves primarily regressive social tendencies; and that the social structure of science, many of its applications and technologies, its modes of defining research problems an designing experiments, its ways of constructing and conferring meanings are not only sexist but also racist, classist, and culturally coercitive"[Harding 1986, p.9].»[57]

Taking these cultural tendencies of our time into account, one realizes the revolutionary potential of the social constructivist account of dynamics of scientific research. For this account has two properties, turning social constructivism into a catalyst for anti-realist (or anti-Galilean) postmodern trends:

First, social constructivist studies of dynamics of research are sometimes very detailed, whereby they appear particularly plausible; secondly, the pretensions of physics as a source of knowledge are clearly undermined in these studies. In Franklin's words:

«Social constructivists imply, however much they may disclaim it, that science does not provide us with knowledge. [...]

The reason the constructivist view is a serious challenge is that their work looks at the actual practice of science. To readers without an adequate science background or knowledge of the particular episodes discussed, the accounts offered by social constructivists may appear persuasive and convincing.»[58]

For the sake of fairness one should admit that the studies of the social constructivists appear attractive not only to laymen but also to some specialists of philosophy of physics. In an early review of Pickering's book *Constructing Quarks* Torretti wrote e.g. the following:

«There is cause enough to celebrate the publication of this book, in which the development of HEP is reported briefly, clearly and entertainingly. [...]. Pickering's book is useful in many regards. I predict that Pickering's work will provoke a violent reaction among the epistemologists of the school,

---

[57] FRANKLIN (1999) 3-4.
[58] Ibidem 4; 6.

which calls itself as realistic. Rather than entertaining us with their fantasies, I hope that they undertake other, no less intelligent and detailed studies of real physics to refute it.»[59]

To sum up it can be said that nowadays the social constructivist view of physics represents the main challenge for the classical (Galilean-Realistic) worldview on physics. Therefore, it is consistent that the final considerations of the work *Constructing Quarks* are dedicated to the divergence between the classical-realistic ideas –that are represented at this point by Polkinghorne– and the constructivist approach. Because here the whole thing is at stake: either nature is rationally structured, and we can recognize this rational order gradually, or the physical theories are bare constructions projecting rationality on the world, which the world actually does not possess at all.

In the latter case Pickering would be right when claiming that «there is no obligation upon anyone framing a view of the world to take account of what twentieth-century science has to say»[60]. And the postmodern humanist rejection of the cognitive claim of physics would thus be justified. But in the first case, one would have to admit that today no philosophy can be advocated that takes no account of the results of physics.

This confrontation is so important for the further development of our culture that it is justified to compare the realist and the constructivist analysis of the dynamics of physical research in detail. In the following chapters I will therefore try to carry out such a comparison with the help of the impulses from astroparticle physics. But first it is important to summarize the recent debate between realists and social constructionists in the field of physics. This is the goal of the next section.

5. *Realistic and Constructivist Case Studies*

Realists and social constructivists have so far focused mainly on analyzing certain episodes of the history of physics. Regrettably, altogether only a few episodes of this history were taken into account, and even fewer cases were investigated from both perspectives (that is realist and constructivist). Franklin mentions four particularly detailed parallel analyses:

«[...] (1) the discovery of weak-neutral currents, Galison (1987) and Pickering (1984b); (2) early attempts to detect gravity waves, Collins (1985, 1994) and Franklin (1994); (3) the solar neutrino problem, Pinch (1986) and

---

[59] TORRETTI (2004) 231-238. (My translation).
[60] PICKERING (1984) 413.

Shapere (1982); and (4) atomic parity violation experiments, Pickering (1984a) and Franklin (1990, chapter 8)»[61].

Franklin, Galison and Shapere represent in these studies realistic positions, while Collins, Pickering and Pinch have adopted the constructivist perspective. By analyzing the debate about the gravitational wave experiments of Weber, Harry Collins tried e.g. to show that the rejection (or acceptance) of the result of an experiment can only be reached by negotiating between specialists, motivated by social interests (career, prestige, benefits, etc.). Collins argues that this is the only way the loop he calls the «experimenter's regress» can be broken, and not by rational arguments and empirical evidence alone[62]. The «experimenter's regress» is due to the experimentalists (understandably) only wanting to accept the results of well-functioning experimental arrangements as correct, but well-functioning experimental arrangements are just the ones delivering correct results. It therefore seems to be no rational way to decide which results we should accept and which not. The researchers must negotiate. Franklin argued in his analysis of

---

[61] FRANKLIN (1999) 10, footnote 17.

[62] Collins's view on the weight of rational grounds and empirical evidence in the dynamics of scientific research seems to have been to some extent modified in his later works "Gravity's Ghost" (2011) and "Gravity's Shadow" (2004). In these works one finds often passages reminiscent of a realistic account of both the physical sources and the devices (See for example the description of working interferometers [p.13-14] and the description of possible sources of gravitational waves [p.24-25] in "Gravity's Ghost").

But there subsist in these books a number of constructivist claims. In some passages Collins seems to insinuate that the phenomena at the frontiers of science are seen or are not seen depending of the actors' interests and agendas. Take, for example, this sentence: «if the old cheap technology really could see the waves, then there would be no need for the new, so the credibility of the old cheap technology had to be destroyed» Collins (2011) p.8.

Some passages are ambiguous: they refer to the rational grounds for taking some decisions, but they point to political and other interests associated with these decisions too. For example here:

«If gravitational waves could be detected for a fraction of the price [...] funding of the big devices would be hard to justify. Therefore, it became a political as well as a scientific necessity to stress that the bars could not do the job that Weber and the Rome Group were claiming for them. On the basis of almost every theory of how these instruments worked, the interferometers were going to be orders of magnitude more sensitive than the bars detectors, and on the basis of almost every theory of the distribution and strength of gravity wave sources in the heavens, only the interferometers had any chance of seeing the waves» Collins (2011) p.12

In this quotation, Collins refers on the one hand to the rational grounds for rejecting the old technology, but on the other hand he insists on the political and/or economical factors for the rejection of old technology. What kind of reasons are the real determinants of the decision? This question remains open.

this case, however, that the rejection of Weber's results were clearly based on rational grounds, and that the endless «experimenter's regress» can be very well broken by mixing rational argumentation and empirical evidence.

But the case that has been investigated and discussed most is undoubtedly the history of the discovery of the weak neutral currents, which was originally described by Pickering (1984) and Galison (1987), and which has later repeatedly been taken up by other authors.

The discovery of the weak neutral currents undoubtedly is a very interesting case because in this episode the researchers were concerned with the revision of established experimental facts, a revision which benefited a certain theoretical proposal, and which could only take place after a very complicated data analysis.

Which facts were revised? The claim of the non-existence of weak neutral currents –a claim that the physicists considered to have already been proven in the sixties– was followed by the assertion of the existence of this currents in the early seventies. Why? In the standard account of the history of particle physics this change is usually justified with reference to the new particle accelerators [Gargamelle (CERN) and Fermilab (HPWF)] which were put into operation in the early seventies. The new particle accelerators revealed new facts that made a revision of the standard claims about the neutral currents necessary.

Andrew Pickering has a different opinion. In his detailed presentation of the case in his book *Constructing Quarks*, he particularly points out the following facts:

–First, even in the experiments in the sixties many events were recorded which were recognized afterwards as neutral weak current events, but were interpreted as a meaningless «noise» at that time.
–A secondly, the discovery of neutral weak currents in the seventies could only be announced after a very complicated (and controversial!) data analysis.

Due to the first fact Pickering derived the consequence that the weak neutral currents could have been discovered much earlier... if there only had been a willingness to do so.

Based on the second fact Pickering states:

«[...] assumptions were made which could be legitimately questioned: one can easily imagine a determined critic taking issue with some or all these assumptions. Moreover, even if all of the assumptions were granted, it remained the case that they were input not to analytic calculation, but to an extremely complex numerical simulation. [...] Thus the skeptic could legitimately accept the input to the calculation but continue to doubt its output.»[63]

---

[63] PICKERING (1984b) 96.

The researchers came to the decision –according to Pickering– not to pursue the possible objections against the discovery of the weak neutral currents. Why? The theory of the electroweak force of Weinberg-Salam implied the existence of such currents and this theory offered both theoretical and experimental physicists great opportunities for the further development of their activities. In his words:

> «Quite simply, particle physicists accepted the existence of the neutral current because they could ply their trade more profitably in a world in which the neutral current was real.»[64]

Peter Galison's analysis of the same episode – encompassing more than hundred and fifty pages of his book *How Experiments End*– presents a very different picture of this decision. Galison's review lets us see the large number of objections that have been evaluated by the different research teams before the existence of neutral currents was accepted. This review also makes it clear that some important researchers were initially indifferent, suspicious or even hostile with regard to the hypothesis of neutral currents until the examination of data from the Gargamelle and the Fermilab experiments forced them to reassess this issue. One of these prominent experimental physicists was David Cline who admitted: «At present I do not see how to make these effects go away»[65].

In summary, it can be said that the investigation of Galison presents the discovery of weak neutral currents as a process, mainly led by rational argumentation based on data, while Pickering understands this discovery as a decision favorable for the researchers, motivated by the attractiveness of the theory of Weinberg-Salam.

Different authors like Bogen, Collins, Franklin, Mayo and Woodward continued discussing this case. In my opinion, there has so far been no clear winner of this debate.

It is in fact difficult to decide between the realist and the constructivist perspective by using such parallel analyses. For there are, as already mentioned before, very few reviews which are carried out in detail by both sides. And therefore it is not reasonable to draw general conclusions.

Considering the case studies that have been examined either only by realists or only by constructivists, the basis for our argumentation is strengthened. On the constructivist side there are, for example, the studies of Collins and Pinch about the confirmation of general relativity theory and the case of cold nuclear fusion. On the realistic site there are, for example, the investigations of Franklin regarding the discovery and subsequent refutation of the existence of the 17-keV neutrinos and the hypothesis of the existence of a fifth physical force as well as

---

[64] Ibidem 87.
[65] GALISON (1987) 235.

the detailed description of the development of particle physics by Falkenburg[66]. In such cases, however, the lack of alternative analysis complicates the comparison between the realist and the constructivist position.

Generally it is to be said that the constructivist approach represents a serious alternative to realism in those cases where it is difficult to distinguish the phenomenon (or the signal sought) from the background noise. In such situations it seems not to be implausible *prima facie* that the fascination or the possible advantages of a theoretical proposal strong influence the work of experimentalists. But to what extent? How significant is the theoretical influence on research? How much invention and how much discovery is part of these research processes?

For me the suitability of historical case studies to answer these questions is limited. The reason for this is that the objects of these studies always are already completed developments. One is therefore already facing the final outcome of the discussions and analysis of the research process under scrutiny. And the only thing, one can still do here is to interpret the process from one perspective or another.

It would therefore be desirable to supplement these historical studies with other methods to resolve the dispute between realism and social constructivism. It would, for example, be interesting to consider whether it is possible to determine possible differences in the expected development of the examined subject by considering ongoing research from a constructivist and realistic perspective. As in that case the actual subsequent evolution of the discipline is used to examine the plausibility of the realistic (or constructivist) approach. In the next section we will then discuss to what extent astroparticle physics is an appropriate field for such an attempt.

6. Impulses for the Discussion between Realism and Social Constructivism from Astroparticle Physics

Finally we have reached the point where it becomes obvious what impulse our discussion can get from astroparticle physics[67]. Because what we need to make the debate between realism and constructivism go one step further beyond the historical case studies, is to investigate the development of a research line that is still going on, and whose current dynamics and possible future development can be different, depending on whether these dynamics are understood in accordance with the realistic or the social constructivist approach. In the ideal case such an investigation could make it possible to examine the predictive capacity of the

---

[66] FALKENBURG, B. (2007)
[67] See FALKENBURG - RHODE (2007) for seminal work about the philosophical aspects of astroparticle physics.

two perspectives when comparing the real development of this research line with our present analysis in some years.

Astroparticle physics provides us with a particularly favorable field for such a project, because:

1. The question of the separation signal/background, and more generally the question of data analysis in the field of astroparticle physics is particularly important.
2. The development of astroparticle physics is far from being complete yet.
3. The complexity of the experiments in the field of astroparticle physics is comparable to the examples of particle physics, discussed so far.
4. There is some theoretical pressure to reach certain goals.

Some remarks can be helpful at this point:

To 1: In the last section we used the example of weak neutral currents to show that the constructivist point of view can appear particularly plausible when the studied phenomena are difficult to separate from the background noise. For in such a case it is *prima facie* conceivable that the theoretical expectations or wishes of the researchers involved wittingly or unwittingly influence –perhaps decisively– the methods and results of their analyses. Therefore, it is important to examine a field in which such effects can occur in order to make a fair comparison between realism and constructivism.

To 2: Because we need a research area that is still going on, we should look at a discipline which does not already have a «standard model». But currently there already is such a model in particle physics and also in other disciplines such as cosmology, stellar dynamics and many other fields of physics. However, the situation in astroparticle physics is much more open, as we will see in the next chapter. Many questions concerning the origin of cosmic rays are still open. The models of the main sources of this radiation are currently under construction and even the experiments and the measurement methods are being developed today. These circumstances are almost ideal for the goal of our investigation.

To 3: Social constructivism has repeatedly emphasized that the complexity of the experiments brings great freedom of interpretation of the results, ultimately leading to a plasticity of the experiments in favor of the theoretical expectations. Therefore, it is important for a fair comparison between realism and social constructivism to examine a field in which the experiments are just as complex as the experiments of particle physics which have so far been considered to be the preferred topic of the constructivist studies. The experiments of astroparticle physics, however, possess such a complexity.

To 4: In the third section of this chapter we saw that the main difference between the social constructivist and realistic view is that from a constructivist perspective the dynamics of research mainly is «top-down» determined, while from the realistic viewpoint the phenomena are predominantly «bottom-up» discovered. That is, in the constructivist approach the theoretical constructions decisively determine the development of physics, while in the realistic approach experimental physics has much more autonomy from the new theoretical impulses. This autonomy goes so far that experimental physicists can reject very insistent (and profitable!) theoretical proposals simply because the experiments –which are not as influenceable as the constructivists think– show that nature is different. Therefore, it is important to use a field of physics as an object of our study where at present there is certain theoretical pressure to reach certain goals. We will see that this is exactly the case in astroparticle physics. Many theorists regard astroparticle physics as a field that can provide important information about the existence of dark matter and evidence for the quantum nature of gravity. In particular, the quantum gravity theorists consider this discipline as the crucial field where the long-awaited connection of their speculative research to empirical facts can take place.

What influence do these theoretical expectations have on the present projects of astroparticle physicists? How determining can the proposals of the theoreticians with regard to the future evolution of this field be? We will try to answer these questions in the course of our investigation. But first it is necessary to provide a brief description of the phenomena and experiments of astroparticle physics, which will serve as background for the further considerations. This is the goal of the next chapter.

2. Chapter: Astroparticle Physics: An Overview for Philosophers

1. Introduction

Astroparticle physics is a scientific discipline lying in between astrophysics, particle physics and cosmology. Its research object consists of high-energy particles that reach the Earth from different cosmic sources. Through the study of such particles it is primarily attempted to learn more about their sources and the interstellar and intergalactic medium. In addition, one expects these investigations to provide new insights into the field of particle physics and cosmology. And besides, it is possible to derive important information for building a quantum theory of gravity from the analysis of certain phenomena of this field.

The discovery of the charged cosmic rays by Victor Hess in 1912 is usually regarded as the birth of astroparticle physics. Astroparticle physics consequently is a research area that already goes back one hundred years[68]. Nevertheless, one can not see this discipline as an already largely completed research program that has become a standard model of its examination object, and in which, because of this, only secondary progress is to be expected in the future[69]. On the contrary, there are still large gaps in our understanding of the cosmic rays. To fill these gaps, a new generation of telescopes, satellites and other experimental arrangements has been developed since the 90ties which provided for a sharp quality improvement in data collecting. Thanks to new technologies and research facilities, an atmosphere of optimism currently prevails in the field of astroparticle physics

All these circumstances make astroparticle physics an ideal field for the analysis of the interactions between theories and experiments in physics, if one does not want to put past episodes of the history of this science in the centre of the analysis. But before we look into the discussion of some details of the dynamics experiment-theory, it is important to get an overview of the entire field of research. The goal of this chapter is to supply such an overview which is to serve as background for the following philosophical considerations. For this purpose the chapter is divided into the following sections:

At first [2 Section], some of the most important phenomena that examine the astroparticle physics are presented. Immediately after that [3 Section] various experiments and techniques are briefly outlined, which are currently being used for the exploration of the described phenomena. Then follows [4 Section] a sketch of those theoretical models that are currently used for explaining the phenomena described in the second section. In addition, some results expected

---

[68] About the history of astroparticle physics see CIRKEL-BARTELT (2008).
[69] If the possibility of unpredictable revolutions is left aside.

by certain groups of theorists from the astroparticle physics research are described. And finally [5 Section] first notes for our analysis of the controversy realism/constructivism are derived from the discussions of the previous sections.

2. Phenomena of Astroparticle Physics

In a first approach to our subject, one could say that the phenomenon studied by the astroparticle physics are the cosmic rays, that is, the flux of cosmic particles that reach the Earth from space. This characterization, however, needs to be qualified in several aspects. At first, it must be pointed out that three very different types of particles are examined in this research field. In addition, the study of these particles has led to the discovery of a large number of physical effects. Therefore, it is not appropriate to speak of «the phenomenon» of cosmic rays (in the singular form). Instead, we will try to list the most important phenomena that are so far discovered in experiments of astroparticle physics on the next pages.

The first phenomenon that should be mentioned is the average stable composition of cosmic rays. The cosmic radiation above 1 GeV consists of about 85% protons, 12% helium nuclei, 2% leptons and 1% heavy atomic nuclei[70]. A small number of neutrinos and gamma rays are also included in this composition. The dominance of the charged cosmic rays (mainly protons and helium nuclei) explains that for a long time, «cosmic rays» and «charged cosmic rays» were considered as synonymous terms. But the great importance of the gamma rays and neutrinos with regard to exploring the sources of cosmic radiation has been intensively getting into the consciousness of the «cosmic rays community». Indeed, today many of the most expensive and most complex research facilities of the astroparticle physics are dedicated to exploring precisely these two components of cosmic radiation. Therefore, it is appropriate to divide this section into three subsections that summarize the characteristics of the charged cosmic rays, neutrinos and gamma photons.

But before we start with explaining the characteristics of these three kinds of astroparticles components, it is interesting to emphasize one aspect common to all types of cosmic rays, namely the following: the cosmic radiation consists in part of particles that possess an extremely high energy (in the range of TeV and beyond). It makes sense to emphasize this point because one of the major challenges for theoretical models of astroparticle physics is to explain, which natural processes may be responsible for the acquisition of such enormous energies by the particles −an energy that partially exceeds the maximum energy of the particles in the largest particle accelerator of the Earth by several orders of magnitude.

---

[70] Data taken from BACKES (2008) 4.

2.1 Phenomenology of Charged Cosmic Rays

As just mentioned, the cosmic rays consist mainly of charged particles, namely of approximately 98% atomic nuclei and 2% leptons. With a few exceptions the proportion of different elements in the cosmic radiation is very similar to the average frequency distribution of the elements in the galaxy[71]. No noticeable temporal variability of this composition was observed. The following diagram shows the frequency of some elements in the cosmic radiation measured during two balloon experiments[72]:

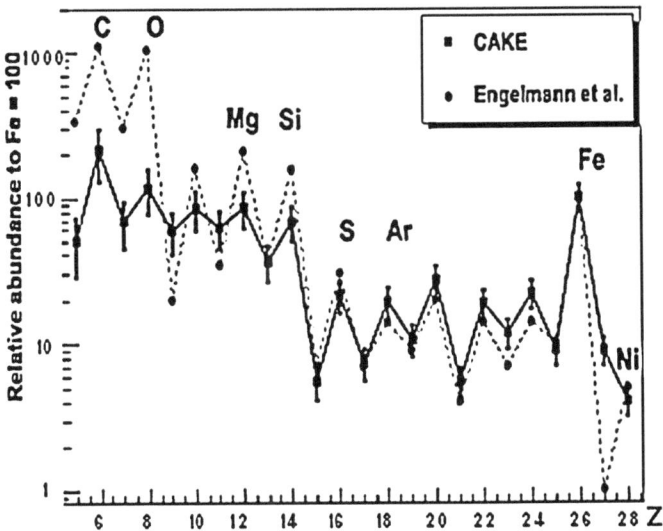

An important feature of the charged particles flux is its isotropy. This phenomenon is explained by the fact that the charged particles are deflected both by intergalactic as well as galactic magnetic fields. And thus they reach the Earth from random directions [see previous image]. Therefore, it is generally not possible to identify the sources of these particles.

Further important phenomena can be inferred from the analysis of the spectrum of the charged cosmic rays. The spectrum shows the particle flux $F=dN/dE$ as a function of the particle energy E. It has the following form[73]:

---

[71] See for example STANEV (2004) 102.
[72] CECCHINI, S. et al. (2009). With permission of the authors.
[73] Credit: Julia Becker

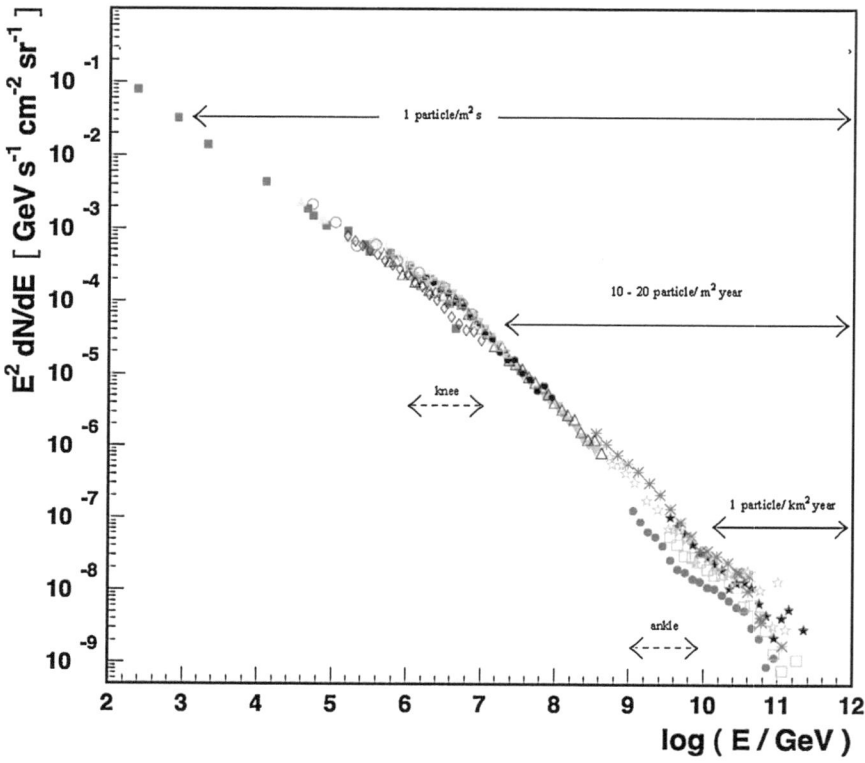

This figure shows the spectrum of the cosmic radiation starting with energy of 100 GeV and weighted with E^2. (The different symbols refer to the experiments, from which the partial measurements of the spectrum originate). The main features of this figure are as follows:

(1) The spectrum can be divided into four parts.
(2) The first three parts can be characterized by a function with the form $F \propto E^\gamma$ (i.e. a power law). The difference between the three parts is shown in the different spectral indices γ:

$\gamma = -2{,}67 \quad E < 10^{6,4} \, GeV$ [section prior to the «knee»]

$\gamma = -3{,}10 \quad 10^{6,4} \, eV < E < 10^{9,5} \, GeV$ [section between the «knee» and the «ankle»]

$\gamma = -2{,}75 \quad 10^{9,5} \, MeV < E$ [section beyond the «ankle»]

(3) The fourth section of the spectrum shows a sharp drop in the particle flux at an energy of approximately $10^{11} GeV$.

In the next section we will take a closer look on the experiments and research facilities involved in measuring this spectrum. And in the fourth section we discuss the models, with the help of whose one tries to explain the mentioned features of the spectrum of the charged cosmic rays. But for now all these details can be left aside, in order to complete the presentation of the phenomena that serve as *explananda* for the theoretical models of astroparticle physics.

2.2 Phenomenology of Gamma-Rays

The gamma-radiation is a form of electromagnetic radiation which consists of photons with energies above 200 KeV. In contrast to the charged cosmic radiation, the gamma-rays are not deflected by magnetic fields. Therefore, it is possible to identify the sources of these high-energy photons. The identification of the sources allows a much more nuanced picture of the processes that are associated with the gamma-part of the cosmic radiation. The main phenomena arising from the investigation of this part of the cosmic radiation are as follows: (1) the link between the gamma-radiation and certain types of cosmic objects; (2) the concrete sky distribution of the sources of gamma-rays; (3) the properties of the spectrum of each source and each kind of sources.

Now we briefly summarize these phenomena. An overview of the types of cosmic gamma sources and the distribution of such objects in the sky can be obtained from a current catalogue of gamma-sources[74]:

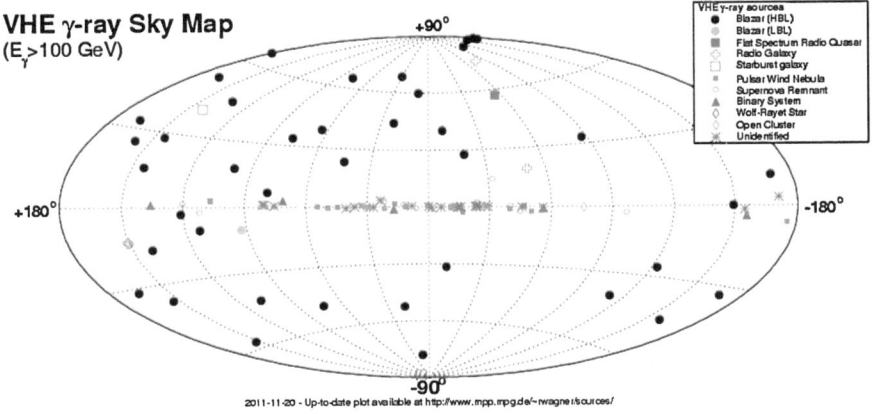

---

[74] Credit: Robert Wagner. Source: http://www.mppmu.mpg.de/~rwagner/sources/

All kinds of sources of gamma-rays –with the exception of the gamma ray bursts which will be dealt with later– are shown in this figure. The illustration uses the galactic coordinate system (i.e., the x-axis is defined by the plane of the galaxy). It can be seen in the figure that the gamma radiation comes from both, galactic and extragalactic sources. The galactic sources are concentrated mainly near the plane of the galaxy, and particularly in the direction of the galactic center. By contrast, extragalactic sources are located in all directions. The kinds of gamma sources represented above are as follows:

(a) Supernovae Remnants [SNRs].

(b) Pulsar Wind Nebulae [PWN].

(c) X-ray Binaries [XRBs].

(d) Open Stars-Clusters.

(e) Unidentified Galactic Sources.

(f) Active Galactic Nuclei [AGN].

A short description of these kinds of sources can be here useful[75]

(a) Supernovae Remnants [SNRs]:
Supernovae remnants are the shell-like remnants of stellar explosions (supernovae), which typically expand for a period of about 1000 years in space with high speeds [at the beginning of up to 10.000 km/s]. The spectrum of gamma-rays was measured in some SNRs. In a typical case –source RXJ0852.0-4622– the spectrum follows a power law with two spectral indices up to an energy level of 20 TeV[76]. Above this energy level the gamma-flux decreases very quickly [cut-off]. There appears to be a correlation between the emission of gamma-radiation and X-rays in objects of this kind[77].

(b) Pulsar Wind Nebulae [PWN].
Pulsar wind nebulae, like the SNRs, are also remnants of Supernovae. The peculiarity of such objects is, however, that a neutron star is formed within the gas cloud of the explosion. Since the angular momentum of the original star remains and the neutron star only has a diameter of few kilometers, this star rotates very fast (within seconds or even faster). This creates a strong magnetic field and charged particles are accelerated along the axis of this field. Due to the acceleration of charged particles, the pulsar emits electromagnetic radiation, even within the range of the gamma-radiation. The spectrum of gamma-radiation, on the TeV

---

[75] For additional information, see e.g. HINTON - HOFMANN (2009).
[76] HINTON - HOFMANN (2009) 539.
[77] Ibidem.

energy range, can be described, as usual, by a power law. In the case of the Crab Nebula –the first source that was detected in the field of TeV gamma rays – the detected spectrum looks like this[78]:

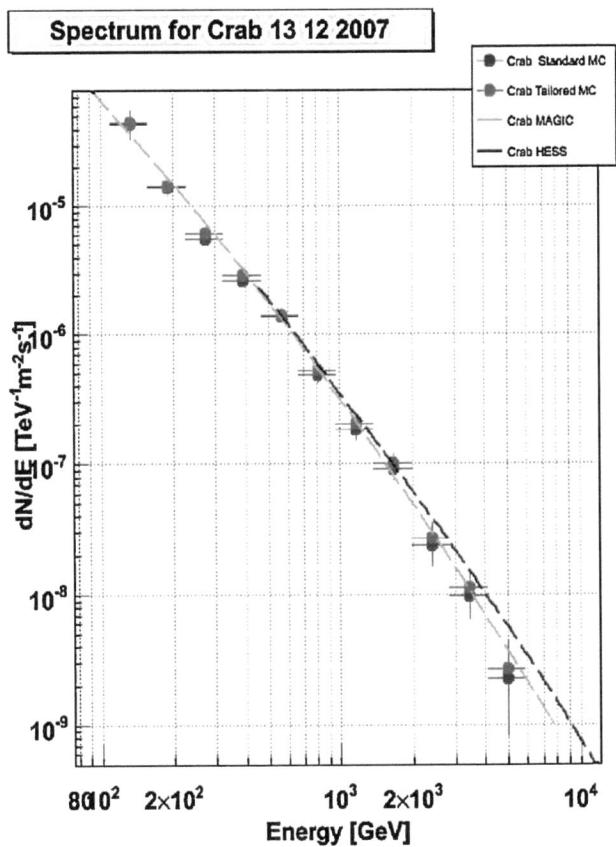

However, considering that part of the gamma spectrum in the low-energy range of about 1-5 GeV –which is feasible with the help of satellite experiments such as EGRET or Fermi– one find pulsating signals[79]. Recently the MAGIC Collaboration reported that in the case of the Crab Nebula such pulsating signals are observed in the range of about 25-60 GeV. The gamma flux in this area is much

---

[78] Source: DOERT (2009) 63.
[79] For additional information, see e.g. WELTEVREDE et al. (2009).

weaker than one would expect from an extrapolation of the satellite data. One therefore assumes that between 5 and 25 GeV there has to be a spectral cut off[80].

(c) X-Ray Binaries [XRBs]:

X-ray binaries are binary stars, consisting of a giant star and a neutron star (or a black hole). These systems emit both X-rays and gamma-rays, which is possibly related to the transfer of matter from the giant star to the compact companion. The flux of radiation in these systems often shows a periodicity on the timescale of a few days, which is associated with the orbital period of the system. In the case of LS 5039, for example, the HESS collaboration measured the following flux variations in the TeV emission[81]:

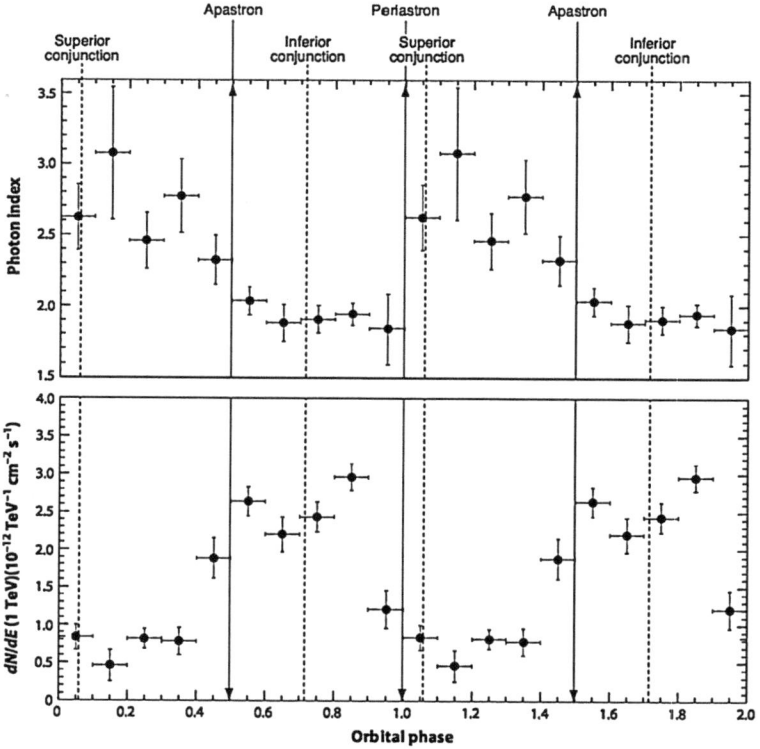

---

[80] See MAGIC-COLLABORATION (2008).

[81] AHARONIAN et al. (2006) 743-49. (With permission).

Also interesting is the phenomenon of unique (or irregular) Gamma flares observed in some X-ray binaries[82].

(d) Open Clusters:
In recent years, it has been confirmed that very energetic gamma-rays can be observed in the direction of open star clusters[83]. The radiation does not appear to come from point sources but from large regions. The spectrum of radiation follows a power law and one cannot detect any significant variability in particle flux. The gamma-emitting open star clusters are regions of intense star formation. One finds Wolf-Rayet stars and binary star systems with a Wolf-Rayet component. The Wolf-Rayet stars are extremely massive stars (up to ~ 50 solar masses) which emit non-thermal X-rays, and they can presumably be sources of gamma rays. It is possible that the observed gamma-radiation of the open star clusters originates from Wolf-Rayet stars. But the extension of the emitting region indicates that gamma rays originate from global stellar wind currents inside the star clusters.

(e) Unidentified Galactic Sources
There are a number of sources of high energy gamma-rays, which have not been observed so far in any other wavelength. Aharonian et al. characterize such objects as follows:

«[These VHE gamma-ray sources] are all extended objects with angular sizes ranging from approximately 3 to 18 arc minutes, lying close to the Galactic plane (suggesting they are located within the Galaxy). In each case, the spectrum of the sources in the TeV energy range can be characterized as a power-law with a differential spectral index in the range of 2.1 to 2.5. The general characteristics of these sources –spectra, size and position– are similar to previously identified galactic VHE sources (e.g. PWNe), however, since these sources have so far no counterpart in lower-energy wavebands, further multi-wavelength study is required to understand the emission mechanisms powering them [...]»[84]

(f) Active Galactic Nuclei [AGN]:
Under the designation «active galactic nuclei» a number of extragalactic sources is grouped, which exhibit very different characteristics from one another: Some are radio-loud and other radio-quiet, some are optically strong and other optically weak, some show broad spectral lines and other narrow spectral lines, etc. However, they all are compiled in a single class, because, according to a plausi-

---

[82] For example, the flare in the system CygX-1, observed by MAGIC. See ALBERT et al. (2007) L51-54.
[83] See for example AHARONIAN et al. (2007a).
[84] AHARONIAN et al. (2007b).

ble model[85] developed in the nineties, most of the differences are due to the angle from which we look at the different AGNs. According to this model, an active galaxy is simply a galaxy, with a central super massive black hole currently in a phase of strong accretion. Vertical to the accretion disk, two jets are formed, in which relativistic particles that cannot be subsumed into the black hole are radiated out of the galaxy. The main structure of an AGN (and the kind of object that is observed depending upon the sight angle) can be seen in this figure/diagram[86]:

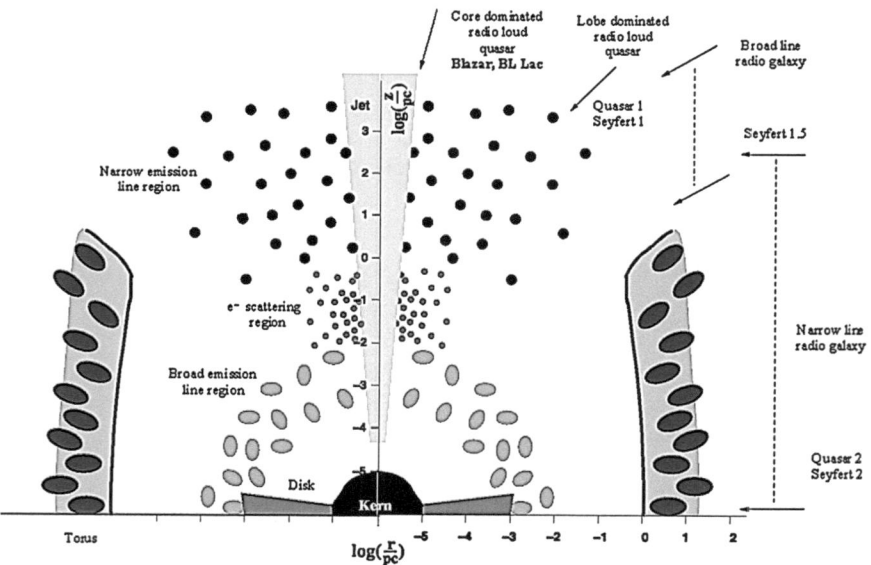

In those cases where the axis of the jet is directed toward the Earth, one speaks about a «blazar».

The AGNs and in particular the blazars are sources of high-energy gamma-rays. Some phenomena that were revealed by investigating the AGN are as follows:

- The sources of gamma-rays appear to be point-like with the resolution of current telescopes.
- In the TeV range the AGNs show a typical power law spectrum. The flux of gamma photons, on the other hand, increases according to the energy in the GeV range[87].

---

[85] See PADOVANI - URRY (1992) 449.
[86] Credit: Christian Zier
[87] See ABDO et al. (2009).

- Many AGNs show a highly variable flux of minutes to days on the timescale. Occasionally it has been observed that the gamma ray flux of an AGN doubles within a few minutes.
- But in about half of the AGNs visible in the TeV range, no signs of variability were observed, yet.
- Usually, the gamma flares of the AGNs are linked to an intensified emission of X-rays. But in the case of the source 1ES 1959 +650 a so-called «orphan flare» was observed in 2002 (i.e., a flare that took place only in the gamma range)[88].
- It is possible to observe gamma-rays from very distant AGNs. Even the quasar 3C279 —located 5,000 million light-years away — has recently been observed with the MAGIC telescope[89].

The objects mentioned so far are steady sources of gamma-rays. But there are additional short-lived sources with a one-time emission lasting less than 100 seconds. These «flashes» of gamma-radiation are called «gamma-ray bursts» [GRBs]. The distribution of gamma-ray bursts in the sky is isotropic. Moreover, so far no event of this nature has been observed at a distance of less than 10kpc. Both facts suggest that the GRBs are extragalactic objects (or phenomena). There are two types of gamma-ray bursts: short and long bursts. The short bursts last less than 2 seconds, while the long bursts can last for approximately a minute or slightly longer. The existence of two different kinds of gamma ray bursts can be recognized, for example, by depicting the duration of the GRBs included in the catalogue BASET 4B[90]:

---

[88] For additional information, see e.g. THOM (2009).
[89] ALBERT et al. (2008) 1752-1754.
[90] BECKER (2007) 28.

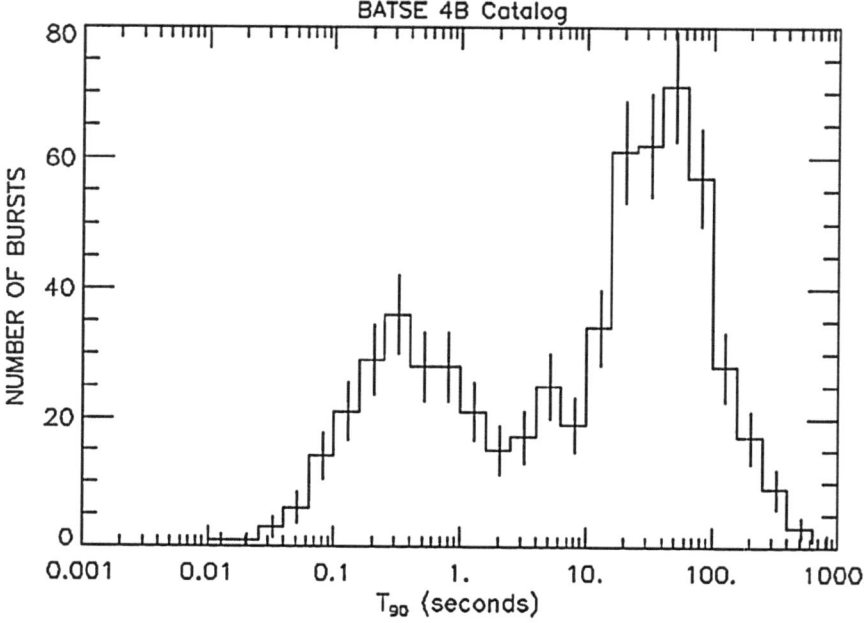

There are other differences between the short and the long gamma-ray bursts. For example, one finds the long bursts mainly in regions in which new stars are produced, while the short bursts tend to be detected in regions with low star formation rates and redshift ($z \sim 0.1$)[91].

Much is still unknown about the nature of these flashes. But at present it is generally supposed that the origin of the short bursts may lie in the fusion of black holes (or neutron stars and black holes), while the long bursts may be associated with hypernovae explosions (i.e. explosions of Wolf-Rayet stars). The Wolf-Rayet stars are extremely massive stars which have a very short life. In the final phase of such a star its shell is pushed off, while the core implodes into a black hole. Around the black hole an accretion disk and two jets, vertical to the disk, are produced. Finally, the star explodes as a hypernovae. Possibly one observes then a long gamma-ray burst, if the jets of the collapsing Wolf-Rayet star point toward the Earth.

For the sake of completeness, it is finally to be mentioned that besides the phenomena and sources of gamma-radiation outlined here, additional kinds of sources (and thus new phenomena) are supposed to exists, which have not yet been confirmed (or at least not clearly). It is commonly assumed e.g. that the

---

[91] BECKER (2007) 34.

Wolf-Rayet stars do not only serve as sources of gamma-radiation at the end of their existence, and that the starburst galaxies and other galaxies and even galaxy clusters are sources of gamma-radiation, too.

2.3 Phenomenology of the cosmic neutrinos

In the last subsection we have seen that the fact that the gamma-radiation is not deflected by magnetic fields, makes a detailed description of many phenomena of the sources of this radiation possible, and equips us with an excellent tool to examine the nature of the astrophysical particle accelerators. Since the neutrinos (just like the gamma-rays) are also not deflected by magnetic fields, one would initially expect that these particles provide another powerful tool for the development of astroparticle physics. In principle, this expectation is justified. But unfortunately, the cross section of neutrinos is so low that the detection of these particles is an extremely difficult enterprise. In addition, neutrinos are also produced in the atmosphere of the Earth as a consequence of slow down processes of the penetrating cosmic particles. The energy range of the atmospheric neutrinos overlaps largely with the energy range of cosmic neutrinos. The consequence of this is, that the atmospheric neutrinos act as a strong background noise that makes the detection of galactic and extragalactic sources of neutrinos almost impossible.

Consequently, it is not surprising that the only sources of astrophysical neutrinos, which have been identified to date, are the sun and the supernova 1987A (in the Large Magellanic Cloud). This means, that the current data are consistent with the possibility that all detected neutrinos (except the solar neutrinos and the neutrinos from the supernova 1987A) are atmospheric neutrinos. A diagram of the observed neutrino spectrum, along with some theoretically expected astrophysical contributions to this spectrum looks like this[92]:

BECKER (2007) 6.

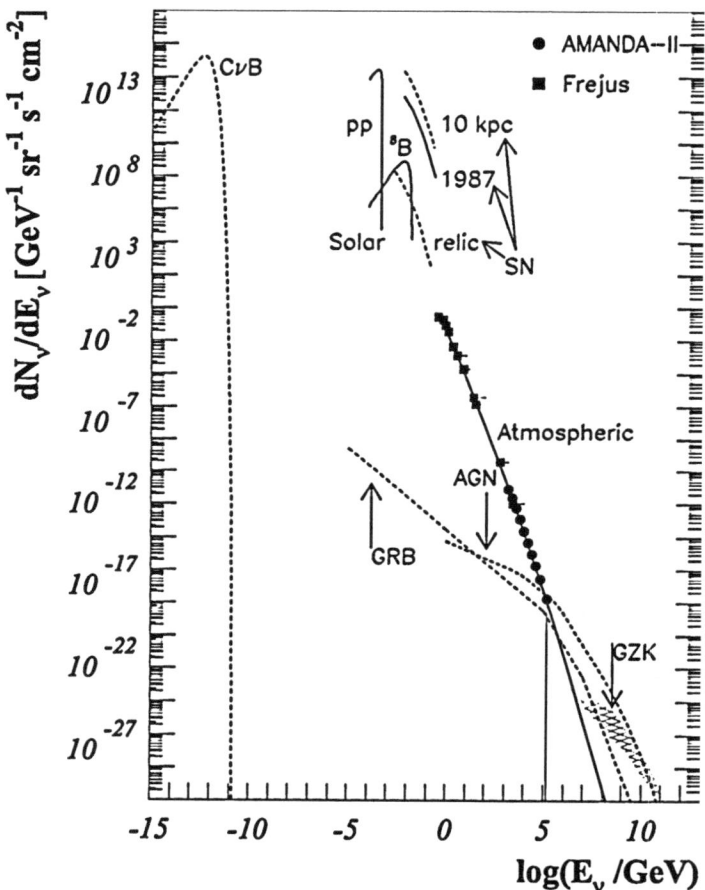

The dashed lines in this plot represent the theoretical expectations of different astrophysical contributions to the neutrino spectrum: CνB is the cosmic neutrino background, which is derived from the standard model of cosmology; GRB represents the expected neutrino contribution of gamma ray bursts, AGN the expected contribution of active galactic nuclei and SN the contribution of supernova neutrinos. The actual detected neutrinos are represented by solid lines. As it can be seen, the atmospheric neutrinos largely overlap the contribution of extragalactic sources. It is only expected for very energetic neutrinos to dominate the number of extragalactic neutrinos.

At present large facilities for the detection of high-energy astrophysical neutrinos are under construction, as reported in the next section. With the current detectors one can only set upper limits for neutrino fluxes which can only be observed from the direction of the potential sources. In addition, it is also possible to search for neutrinos from the direction of gamma-ray bursts. But so far, the number of detected neutrinos that could be associated with such events is very low, and the possibility that only the atmospheric background was observed, cannot be excluded, yet.

2.4 Parenthesis for Philosophers: The Concept of «phenomenon»[93]

On the previous pages we have briefly outlined the main phenomena that have been discovered and investigated by the astroparticle physics so far. In this review, the term «phenomenon» has been used in the sense that it is commonly used by researchers in this field. A phenomenon in this sense is often a type of events that is specified by a number of characterizing properties. (Two event types, for example, are the short and long gamma-ray bursts). Phenomena also are some correlations between different physical parameters –the so-called «empirical laws»– occurring in certain circumstances. (Empirical laws of astroparticle physics are for example the numerous power laws that are characteristic for the spectrum of each source-type). In addition, a concrete realization of an event type or a concrete realization of an empirical law is often also called a «phenomenon». The phenomena of this kind are usually derived as a result of the data analysis of a given data set.

The phenomena in all mentioned meanings –i.e. the abstract event types and empirical laws on the one hand and the concrete realization of these abstract phenomena in concrete cases on the other hand– are entities of a very special nature. They are certainly theoretical entities[94]. But they are largely dependent on the «raw data», which represent no more theory, but empirical ground. And the «processing» of this raw data is realized with the help of theoretical strategies and techniques (such as probability theory, measurement theories and theories about the measuring instruments etc.) which have already been successfully tested in other contexts. Therefore, one can say, that the phenomena are exactly on the border between theory and experiment. In other words, the phenomena can be understood as the intermediary between theories and experimental results. On the one hand they represent the first piece of theoretical knowledge, which

---

[93] See more details about the concept of «phenomenon» in physics in FALKENBURG (forthcoming).
[94] In the case of abstract phenomena the reason for this assertion is clear. But for the concrete realizations of these abstract types the statement is valid too, because these «concrete» phenomena are not something «empirically given», but they are instead the result of a (usually very complex) process of data reduction and data analysis.

can be found by experiments, and on the other hand they are the *explananda* of scientific explanations by theories and models.

If the realistic view of the dynamics of scientific progress is true, the phenomena are based only (or essentially) on the raw data from the experiments and the theoretical ground knowledge (the accepted measurement theories and other well established theories). But should the constructivist view be true, it is to be expected that the data analysis that lead to the constitution of phenomena based on the raw data are effectively influenced by the theoretical hypothesis, in the confirmation of which the experimentalists are particularly interested. Therefore, it is very important for our investigation to analyze how the phenomena of astroparticle physics are derived from the data analysis –a task that will be in the focus of the next two chapters of in this study–.

The understanding of phenomena as mediators between theories and experimental raw data, corresponds at large with the now usual approach of the philosophy of science; an approach which is mainly derived from the influential paper by Bogen and Woodward «Saving the Phenomena»[95]. It should be noted, however, that the examples of phenomena that Bogen and Woodward discuss in their paper, cannot (at least in part) be regarded as phenomena in the sense used here, but rather as theoretical interpretations of phenomena. Bogen and Woodward call e.g. the neutral weak currents a phenomenon that can be explained by the electroweak theory of Weinberg-Salam. However, the phenomenon discovered by the Gargamelle-experiment at CERN and the experiments in the American NAL were not –contrary to the claim of these authors– the neutral weak currents, but the existence of a certain proportion between neutral events and the charged events in their records. A proportion that implies a higher number of neutral events than expected for the case, that all neutral events were only background events. The phenomenon therefore is an empirical law (the proportion of «neutral events/loaded events»), whereas the neutral weak currents represent a theoretical interpretation (or explanation) of this phenomenon.

This difference between a phenomenon and a theoretical interpretation of the phenomenon is often neglected in the colloquial language of researchers. The cause of such careless formulations is, that although the experts usually know what has actually emerged as a phenomenon of an experiment (or of an observation-campaign), they usually look at the phenomena from the perspective of their hypothetical explanations. This way of formulating may at first cause some confusion about the purposes of an investigation of the interaction between theory and empirical data, but it does not represents an insurmountable difficulty, because, if necessary, the phenomena can be formulated independently of these theoretical interpretations/explanations.

---

[95] BOGEN, J. UND WOODWARD, J. (1988), «Saving the phenomena»: *The Philosophical Review* XCVII N°3 (1988) S.303-352.

In this section I have tried to describe the most relevant phenomena of astroparticle physics for the most part independently from theoretical interpretations. But strictly speaking, not all the descriptions in this section have been free of such interpretations. For example, for the sake of clarity were added in the presentation of the phenomena associated with the different types of sources of gamma-radiation, from time to time hypothetical assumptions about the nature of these sources. Basically we can say that the phenomena of astroparticle physics are the proportion of the different kinds of particles in the cosmic radiation, the form of the sources (point-like or extended), the spectra of the sources and the possible variations of particle fluxes of the sources. Everything else is theoretical interpretation.

3. Experiments, Observatories and Techniques in Astroparticle Physics

When a cosmic particle reaches the Earth's atmosphere diverse interactions can take place depending on the type of particle. This fact makes it possible (or even necessary) to use different methods of detection:

If the particle is a neutrino, it will very probably cross the Earth without any interaction. Only occasionally a weak interaction between a neutrino and an atomic nucleus take place in the atmosphere or in the Earth itself. In this case, charged leptons are originated, which can be detected by large detectors in water, ice or other substances. (Due to the minimal cross section of the neutrinos, the detectors must be really large, which implies a number of technical difficulties).

If the particle is a gamma photon, it will not reach the Earth's surface, due to the interaction with the atmospheric molecules. The gamma-rays must therefore be detected either directly by satellites and balloon experiments outside the atmosphere or indirectly by air Cherenkov telescopes and experiments. These telescopes use the fact that an incoming gamma photon causes an avalanche of secondary particles in the Earth's atmosphere –a so-called «particle shower» or «particle air shower»–, in which Cherenkov light is produced.

In the case of the charged cosmic rays, many interactions of the incoming particles with the atmospheric molecules take place, and they produce particle showers, too. Therefore, it is possible to use satellites and balloons for the direct detection and ground- experiments based on Cherenkov- light for the indirect detection of these particles. Similarities and differences between the gamma-induced and the hadron-induced particle air showers can be illustrated with a schematic representation:

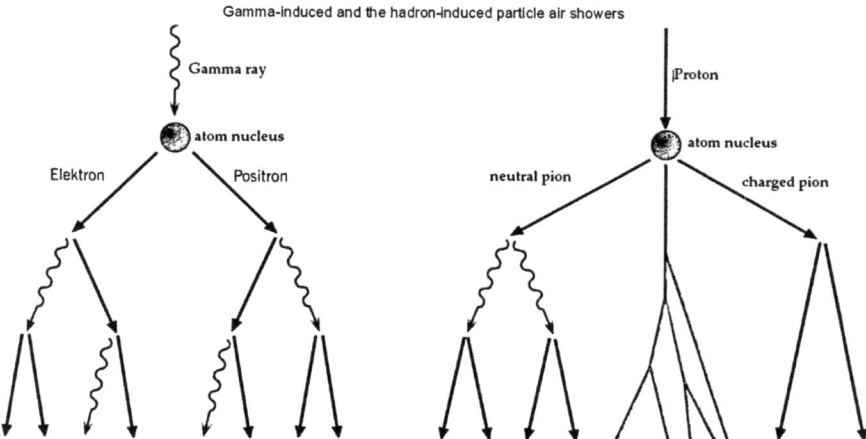
Gamma-induced and the hadron-induced particle air showers

For the detection of charged cosmic rays detectors have also been developed, which can detect the UV light that is produced when charged particles of the particle showers interact with nitrogen molecules in the atmosphere. (Those are the so-called «fluorescence experiments»).

In view of such variety of experiments and proof techniques, this section will be divided into six subsections, which outline (1) the satellite detectors (2) the balloon experiments (3) the fluorescence experiments (4) the air shower ground experiments (5) the Cherenkov-telescopes (6) the underground detectors. In each subsection, some general aspects of the applied technique are explained at first and then some concrete experiments are briefly described as examples.

For lack of space I can (unfortunately) not mention many remarkable experiments and research facilities. I therefore recommend the website http://www.mpi-hd.mpg.de/hfm/CosmicRay/CosmicRaySites.html of the Max Planck Institute for Nuclear Physics in Heidelberg. This website −created by Konrad Bernlöhr− contains a collection of links to the websites of most experiments in astroparticle physics, where much more details about the techniques used and the progress of research can be found.

Finally, a seventh Subsection summarizes some comments for our comparison of the realistic and the constructivist view of the dynamics of scientific progress that can be gained from a consideration of the astroparticle physics experiments.

3.1 The Satellite Detectors

The satellite observations are particularly appropriate for the direct detection of charged cosmic rays and gamma-rays, because the satellites orbit the Earth at an altitude of several hundred kilometers, hence in regions where no atmospherically caused disturbance of the measurements can take place. The main difficulty of this procedure is however that the launching rockets can place only small detectors in space. Since the flux of cosmic particles strongly decreases with energy, this means that the satellite experiments can observe the particle flux only up to energies of about 100 GeV. The TeV region of the spectrum of cosmic particles cannot (or barely) be reached by satellite detectors.

The components of the diverse satellite experiments vary not only with regard to the technological development, but particularly in connection with the characteristics of the particles to be detected. There are however components, which practically belong to each (current) satellite detector. These include in particular:

(1) A calorimeter for measuring the energy of each particle.

(2) An «anticoincidence» system to suppress the background events.

(3) A trigger system to decide which events are stored and processed.

Additionally, the satellites that examine the charged cosmic rays are doted with (4) a magnet spectrometer (for the identification of the kinds of particle) too, and the gamma satellites are doted with (5) a tracker system (for the determination of the flight direction of the photons) too. As an example for these two kinds of satellite experiments we will deal with some more details about the concept of the experiments PAMELA and FERMI LAT:

3.1.1. PAMELA [Payload for Antimatter Matter Exploration and Light-Nuclei Astrophysics]

PAMELA is a satellite experiment, which was developed for the investigation of the charged cosmic radiation (and especially the anti-matter portion of this radiation). The PAMELA instrument was launched on June 15th 2006 as part of the Russian satellite Resurs-DK1. This satellite orbits the Earth at a height between 350 and 600 km. The essential components of the instrument can be seen in the picture[96]:

---

[96] Source: CASOLINO et al. (2009). With permission

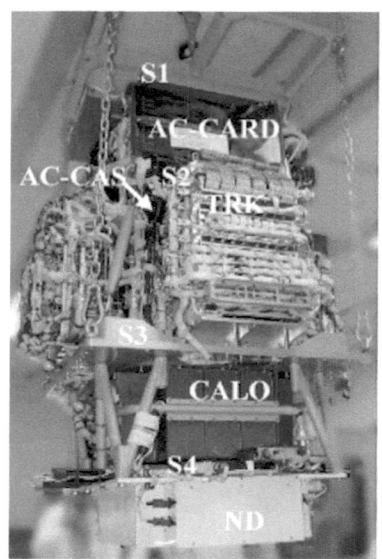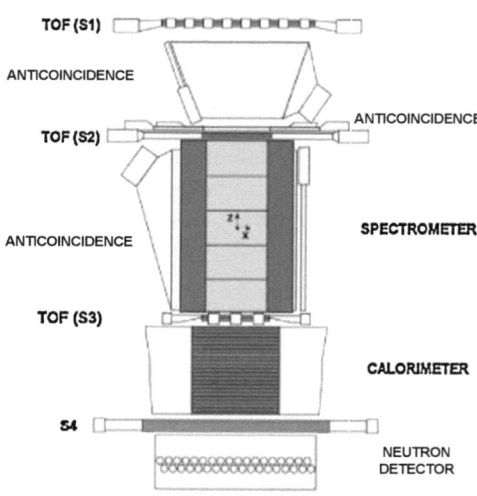

On the official website of the PAMELA experiment[97], we find the following brief description of the diverse components:

«A *Magnetic Spectrometer*, based on a neodymion-iron-boron permanent magnet and a precision tracking system.

Charged particles passing through the instrument aperture are deflected by the magnetic field. The resulting curved trajectories depend on the electric charge-sign and magnetic rigidity (Momentum/ charge) of the particles. The sign of the electric charge is determined with a very high confidence degree, and the momentum of the particle is measured up to the highest energies for which a useful statistics for flux of rare particles, as antiprotons and positrons, can be collected;

A *Sampling Imaging Calorimeter*, is used to interleave pairs of orthogonal ministrip silicon sensor planes are interleaved with tungsten absorber plates. The energy released by the interacting particle (and indeed of the extra energy in case of an annihilation event) are measured and the interaction topology of the particle inside the calorimeter is reconstructed. The imaging calorimeter is the primary means of distinguishing electrons and positrons from antiproton and proton with the same charge sign and momentum, giving a

---

[97] http://pamela.roma2.infn.it/index.php.

separation factor on the order of 10^5. Observing the annihilation pattern of antiprotons and possible antinuclei, a confirmation of their identity on an "event by event" basis is obtained;

A precise *Time of Flight System*, using plastic scintillation detectors, measures the charge of the incident particles, the versus of their flight, for albedo rejection, and the velocity of low energy particles, to determining, combined with the rigidity, the mass. The coincidence of TOF scintillators provides flight reprogrammable fast trigger to initiating event digitisation and read out.

An *Anticoincidence System*, based on plastic scintillation counters placed at sides and around top of the magnet and at sides of the instrument between the first and the second counter layer of TOF flags events in which particles may have entered through the magnet or have produced in pairs with only one traversing the detector.

A *Neutron Detector* composed by 3He counters and polyethylene neutron moderators, to distinguishing primary protons and electrons at energies 1011-1013 eV with a rejection factor of about 10-4.

A *Bottom Scintillator* counter for independently triggering the neutron detector.»

At this point the following three remarks can be interesting for the purposes of this study:

(1) As usual in current experimental physics, all components of PAMELA were first examined individually, calibrated and tested with the help of simulations. Then the parts were assembled, and further calibrations and tests were made before the device was considered to be functional:

«[...] tests involved first each subsystem separately and subsequently the whole apparatus simulating all interactions with the satellite using an Electronic Ground Support Equipment. Final tests involved cosmic ray acquisitions with muons for a total of about 480 hours. The device was then shipped to TsKB Progress plant, in Samara (Russia), for installation in a pressurized container on board the Resurs-DK satellite for final tests. Also in this case acquisitions with cosmic muons (140 hours) have been performed and have shown the correct functioning of the apparatus, which was then integrated with the pressurized container and the satellite. The detector was then dismounted from the satellite and shipped by air to Baikonur cosmodrome (Kazakstan) where the final integration phase took place in 2006».[98]

---

[98] CASOLINO et al. (2009).

(2) Although it is clear that the construction of the equipment used in the experiment has required a considerable amount on research and technology development, the basic ideas of the various detectors are well known. The principle of the magnetic spectrometer has e.g. been used frequently in the context of particle physics. And the same can be said of the calorimeter and the other components of the experiment. This implies, among other things, that the construction of PAMELA has been accompanied by a certain, already existing technical knowledge about the structure, the calibration methods and the working mode of the various devices. This in turn implies that the instruments have to meet certain objective standards and therefore could not be arbitrarily adjusted to produce desired results.

(3) The installation of diverse trigger systems and an anticoincidence system points unmistakably to the fact that the desired signal can be perceived only if the strong background can be eliminated. This implies that the background simulations and various separation methods must play necessarily an important role in data analysis of the PAMELA files. Consequently, this is a research project by which the constructivists may start raising similar objections against the realism as we have discussed in the first chapter in connection with the neutral weak currents. This encourages our idea that the experiments of astroparticle physics offer a suitable basis for the continuation of the debate between realism and constructivism.

Similar observations can also be stated concerning the other experiments, which are to be discussed on the following pages. We will evaluate the significance of these facts for our discussion at the end of this section.

3.1.2 FERMI-LAT [Fermi Large Area Telescope]

FERMI is a satellite that was launched on 11 June 2008 to study the gamma-radiation. The satellite includes two scientific instruments: The main instrument is a gamma telescope (Fermi-LAT), supplemented by a gamma-ray burst monitor. The essential components of the LAT instrument are 16 square track-towers, ending in a calorimeter. They are outlined in the following picture [99]:

---

[99] Source:
http://teacherlink.ed.usu.edu/tlnasa/reference/ImagineDVD/Files/imagine/Images/glast/glast_lat_cutaway2.jpg

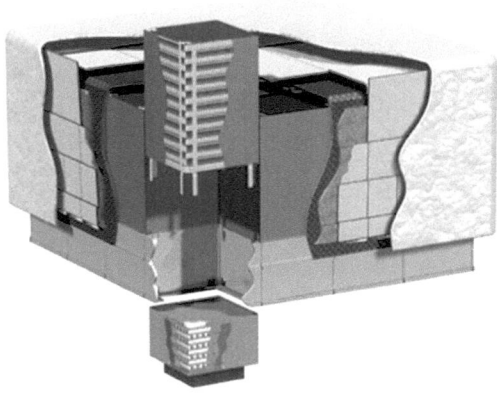

On the website of NASA one finds the following short description of the different components of FERMI LAT:

«Tracker

The Tracker consists of a four-by-four array of tower modules. Each tower module consists of layers of silicon-strip particle tracking detectors interleaved with thin tungsten converter foils. The silicon-strip detectors precisely measure the paths of the electron and positron produced from the initial gamma ray. The pair-conversion signature is also used to help reject the much larger background of cosmic rays.

Calorimeter

The Calorimeter measures the energy of a particle when it is totally absorbed. The LAT Calorimeter is made of a material called cesium iodide that produces flashes of light whose intensity is proportional to the energies of the incoming particle. The Calorimeter also helps to reject cosmic rays, since their pattern of energy deposition is different from that of gamma rays.

Anticoincidence Detector (ACD)

The Anticoincidence Detector is the first line of defense against cosmic rays. It consists of specially formulated plastic tiles that produce flashes of light when hit by charged-particle cosmic rays (but not by gamma rays, which are electronically neutral). The ACD forms a "hat" that fits over the tracker.

Data Acquisition System (DAQ)

The Data Acquisition System is the brain behind the LAT. It collects information from the Tracker, the Calorimeter, and the Anticoincidence Detector and makes the initial distinction between unwanted signals from cosmic rays and real gamma-ray signals to decide which of the signals should be relayed to the ground. This system also does an on-board search for gamma-ray bursts. The DAQ consists of specialized electronics and microprocessors.»[100]

The function mode of the FERMI-LAT can be summarized as follows: when a gamma quantum penetrates into one of the track-towers it can generate a pair of subatomic particles (an electron and a positron). The direction of the incoming gamma quantum is determined by reconstructing the direction of these particles back to their source, using several layers of silicon detectors in the tracker. Subsequently, the calorimeter absorbs and measures the energy of the particles. This detection procedure is supported by an anticoincidence system and a data acquisition system, designed to separate the background events in the detector as far as possible from the relevant signal.

At this point two comments may be useful:

(1) Above it has been pointed out that the construction of PAMELA is based on a certain, already existing technical knowledge about structure, the calibration methods and the working of the various devices. In the case of FERMI-LAT, we can repeat this comment, and we can also emphasize that this connection to the existing technical knowledge was a central request in the planning of the experiment. And so, Atwood et al., for example, wrote:

«[...] detector technologies were chosen that have an extensive history of application in space science and high-energy physics with demonstrated high reliability.»[101]

In addition, a whole set of calibration and test procedures were also applied here –as in the construction of the PAMELA Instrument–:

«[...] relevant test models were built to demonstrate that critical requirements, such as power, efficiency, and detector noise occupancy, could be readily met. [...] these detector-system models, including all subsystems, were studied in accelerator test beams to validate both the design and the Monte Carlo Programs used in the simulations [...].

The modular design of the LAT allowed the construction of a full-scale, fully functional engineering demonstration telescope module for validation of the

---

[100] http://www.nasa.gov/mission_pages/GLAST/spacecraft/index.html
[101] ATWOOD et al. (2009)

design concept at reasonable incremental costs. This test engineering model flew in a high-altitude balloon to demonstrate system level performance in a realistic, harsh background environment [...] and was subjected to an accelerator beam test program [...]. Particle beam tests were also done on spare flight tracker and calorimeter modules.»[102]

Both the application of proven techniques for the production of the components and the use of strict calibration procedures suggest that the «plasticity» of FERMI-LAT concerning a hypothetical adjustment of its function mode in order to produce some results that are desired by theorists, is very low or not existent. That is, if the theoretical trends lead and set the curse of development of the experimental physics –as constructivism suggests–, it seems unlikely that this control is due to a clever «manipulation» of the tools for data collection in experiments as FERMI-LAT or PAMELA (or other experiments in astroparticle physics). The control of the experiments through the theoretical trends should therefore become visible in the data analysis.

(2) The important role of the ACD and the trigger systems of FERMI-LAT once again underscore the serious problem of separation of signal and background with which the experiments in astroparticle physics are confronted. During the observations with FERMI-LAT this background consists of charged cosmic rays –mainly hadrons– and gamma photons that are reflected from the Earth's atmosphere [*albedo*]. Due to the non-removable background it is unavoidable that a complicated process of data analysis precedes each assessment based on the FERMI LAT raw data files. In the next chapter we will therefore pursue –by discussing another experiment–, among other things, the question, to what extent the process of data analysis opens up the possibility of a theoretical guidance of the experiment results.

3.2 The Balloon-Experiments

An alternative to the satellite experiments for the direct investigation of the charged cosmic rays and the gamma-rays is the balloon experiment. A major advantage of balloon experiments is, that they are much less expensive than the satellite experiments. But they have also some disadvantages, such as the fact that the disturbances in the measures due to the atmosphere cannot be eliminated completely, or the fact that the balloons are not suitable for long-term observation, because their flights take only a few days.
In any case, these experiments are performed using devices very similar to those of satellite experiments. For this reason, the examples of the balloon experiments

---

[102] Ibidem.

are presented briefly and without further remarks. As example of the balloon experiments, we now consider briefly the concept of the experiments ATIC and TIGRE:

3.2.1 ATIC [Advanced Thin Ionization Calorimeter]

The ATIC balloon flew for the first time around the Antarctic in 2000. Subsequent flights followed until 2007. The main goal of this experiment was to determine the spectrum of different elements (from H to Fe) of the cosmic radiation. Another goal of the experiment was to determine the energy dependency of the relation H/He in the cosmic radiation.

The essential elements of the ATIC instrument can be seen in the following picture[103]:

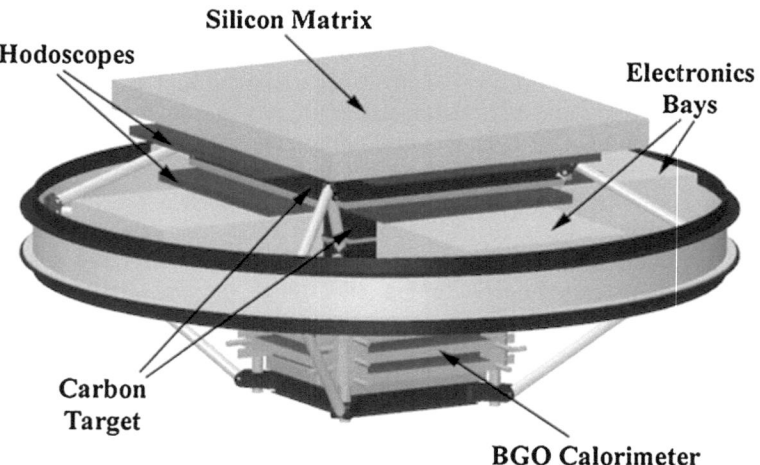

The main component is the calorimeter, which enables to measure the energy of the penetrating cosmic nuclei from H to Fe. Over the calorimeter, there is a three-layer hodoscope and two layers of inert carbon. The hodoscope defines the opening surface of the device and also measures the charge and the direction of the penetrating particles. The carbon layers are inserted with the goal that as many penetrating cosmic ray particles as possible interact with the detector. The top part of the equipment is a silicon matrix detector, which can also measure the

---

[103] Source: http://atic.phys.lsu.edu/Instrument.html

charge of the penetrating particles. The electronic devices for data selection, the energy system and other electronic instruments are placed around the diverse detectors.

3.2.2 TIGRE [Tracking and Imaging Gamma-Ray Experiment].

The «Tracking and Imaging Gamma Ray Experiment» (TIGRE) – Group at the University of California Riverside has been working for several years on the development of a balloon experiment for the observation of cosmic gamma-rays. This experiment is thus still in a developmental stage. In 2007 a prototype of the TIGRE balloons flew for the first time (for two days) and detected approximately 2 million gamma events. As it is the rule in gamma-ray experiments, the main elements of the TIGRE equipment are a tracker and a calorimeter. The essential structure of the device can be seen below[104]:

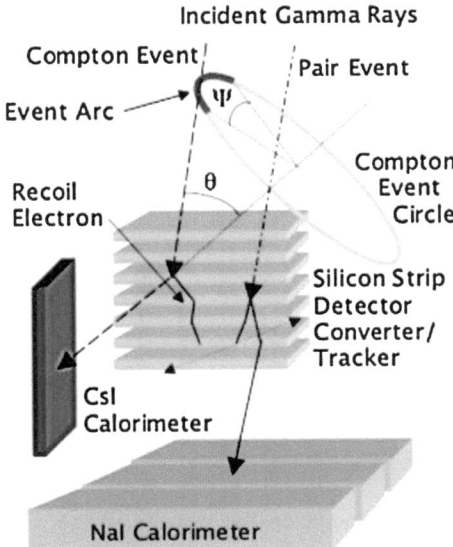

Compton and pair events in the silicon converter/tracker

As the figure shows, the TIGRE - equipment can recognize two different phenomena of the gamma radiation: the production of an electron-positron pair and

---

[104] Credit: Allen Zych

the Compton-scattering of electrons from the tracker. Through the detection of Compton-events, gamma photons between 1 and 10 MeV can be identified. Through the detection of electron-positron events, gamma photons between 10 and 100 MeV can be identified.

3.3 The Fluorescence Experiments

The fluorescence experiments use to investigate cosmic radiation a technique very different from those described up to this point. The basic idea of such experiments is as follows: as mentioned above, the incoming cosmic rays cause an avalanche of secondary particles in the atmosphere, called «particle showers» or «particles of air showers». The passage of these particles through the atmosphere leads to ionization and excitation of air molecules (mainly nitrogen molecules). These molecules then return back to their ground state by emitting UV radiation. And so each particle shower is accompanied by a UV flash, which can be detected by lenses or mirror systems. A schematic representation of this type of detection of cosmic rays is as follows[105]:

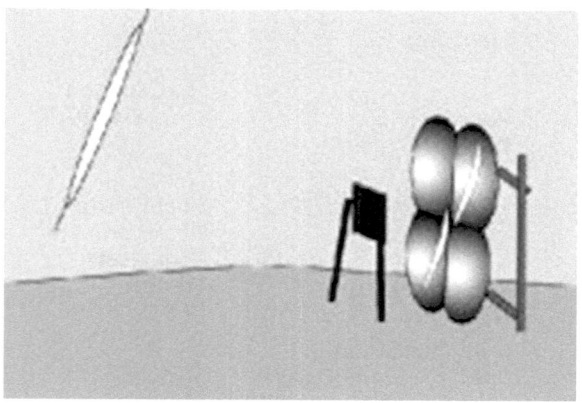

This image, which can be found on the website of the «Telescope Array Project», is accompanied by the following explanation:

> «This figure shows a schematic of a fluorescence air shower detector. The scintillation light is collected using a lens or a mirror and imaged on to a camera located at the focal plane. The camera pixelizes the image and records the time of arrival of light along with the amount of light collected at each pixel element. This technique can be made to work on clear, moonless nights,

---

[105] Source: http://www.telescopearray.org/outreach/history.php

using very fast camera elements to record light flashes of a few microseconds in duration»

It should be noted that the name «fluorescence» for this kind of detection technique is not absolutely correct, because the used phenomenon is not fluorescence, but a kind luminescence. But, to the dismay of opticians, this name has been firmly established among the astroparticle physicists. Examples of fluorescence experiments in astroparticle physics are the «stereo Fly's Eye», «MIA» and «HiRes» experiments. A brief comparative sketch of these experiments follows:

«In [the Fly's Eye] experiment, two fluorescence detectors were positioned 3.5 km apart. The effective energy range extended from 10^{17} eV to 3×10^{19} eV. This experiment utilized optics that resulted in a 5×5 degree pixel size, and the geometry of the event was determined by the intersection of the two event-detector planes, with resulting resolution in Xmax of about 45 gm/cm^2. Both Fly's Eye detectors had mirrors that covered an elevation angle range from 0 to 90 degrees, though FE II had only partial azimuth coverage.

The HiRes Prototype/MIA experiment was a hybrid fluorescence - muon array detector. The fluorescence detector, a fourteen mirror prototype of the current HiRes detector had a 1×1 degree pixel size and was located 3.5 km from the center of the MIA muon array. The muon array sampled a part of muon lateral distribution. The effective area of this "array" was thus set by the muon lateral distribution function and was roughly a 3 km radius centered on MIA. The smaller pixel size of the HiRes detector allowed triggering to lower energies (approaching 5×10^{16} eV) while the effective area of MIA limited the energy reach to just above 10^{18} eV. The Xmax resolution was 45 gm/cm^2. The geometry of the event was determined by using the HiRes event-detector plane and timing information from the fluorescence detector and the MIA ground array.

The HiRes stereo detector consists of two fluorescence detectors separated by 12 km. With 1×1 deg pixel size, the Xmax resolution is 30 gm/cm^2. The large separation between detectors and the limited elevation angle range (3 to 32 degrees) limits the energy range to > 10^{18} eV with the statistical reach approaching 10^{20} eV.»[106]

We can dispense with a detailed description of the detection method of fluorescence flashes, because it is a technique that is closely related to the technique of the Cherenkov Telescopes, which is discussed in detail in the next chapter. At this point it is thus sufficient, to make the following remarks:

---

[106] SOKOLSKY et al. (2005).

(1) The fluorescence detectors are based on founded knowledge in optics, electronics, etc., which were also successfully used for the construction of numerous other devices. This means that –as already mentioned in other types of experiments– the «plasticity» of the detectors regarding an adjustment of its function mode, in order to produce some results that are desired by theorists, is very limited (at most).

(2) The data analysis, on the other hand, turns out to be difficult and not free from uncertain steps[107].

This shows once again that the question of whether the theoretical trends determine the experimental results (at least in crucial spots/places) is to a large extent a question of the correct understanding of the processes of data analysis.

3.4 The Air Shower Ground Experiments

The air shower ground experiments are used mainly to detect charged cosmic rays above an energy of about $10^{5.5}$ GeV. However, not the primary cosmic rays are detected but the secondary particles resulting from interactions of the cosmic rays with atoms and molecules in the Earth's atmosphere.

An advantage of the ground experiments is, that the detector surfaces can be much bigger than the ones of satellite detectors. Thus, observations of the high-energy particles are made possible, while these particles –due to their small number– cannot be observed with satellites. The disadvantage of such experiments is, however, that the calculation of the energy and the track of the primary particles from the indirect measurements that can be implemented at the secondary particles, is a very complex and uncertain task. The calculation of the energy turns out to be particularly difficult, because numerous important parameters to describe the development of a particle shower are only insufficiently known. This is particularly true for the cross section of many interactions within the particle showers caused by the most energetic cosmic rays. After all, the cross section can only be determined by extrapolation in these cases, because the energy of such interactions is not reached in current particle accelerators. For this reason, some ground experiments are designed to determine the energy of the penetrating primary particles with more than just one

---

[107] And so, for example, Zech mentions the following sources of error, which can decisively distort the data analysis of HiRes, if they are not taken into account:
«The main systematic uncertainties that are introduced in the measurement of the ultra-high energy cosmic ray spectrum with the HiRes experiment [include] uncertainties in the phototube calibration, fluorescence yield, "missing energy" correction and aerosol concentration in the atmosphere, the total systematic uncertainty in the measured flux is 31% for each of the two monocular spectrum measurements». ZECH (2005).

method. These experiments are called «hybrid experiments». An important example of this kind of experiments is the «Pierre Auger Experiment», which is therefore briefly outlined below:

3.4.1 The «Pierre Auger Experiment»

The Pierre Auger experiment consists of 1600 detectors for particle showers, positioned 1,5 km apart from each other in the Argentine Pampa (province of Mendoza). This large detector field is flanked by four fluorescence detectors, which can observe (under suitable atmospheric conditions) the UV light flashes associated with the particle showers[108]:

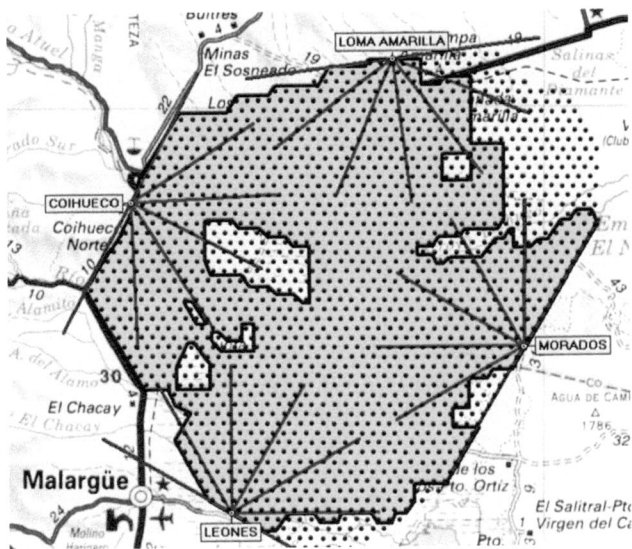

The field covers a total area of approximately 3000 km^2 and is situated at an average altitude of 1,400 m.

On the website of the experiment, we find the following picture and a brief description of the shower detectors[109]:

---

[108] Source: http://www.auger.org/
[109] Source: http://www.auger.org/features/inside_surface_detector.html

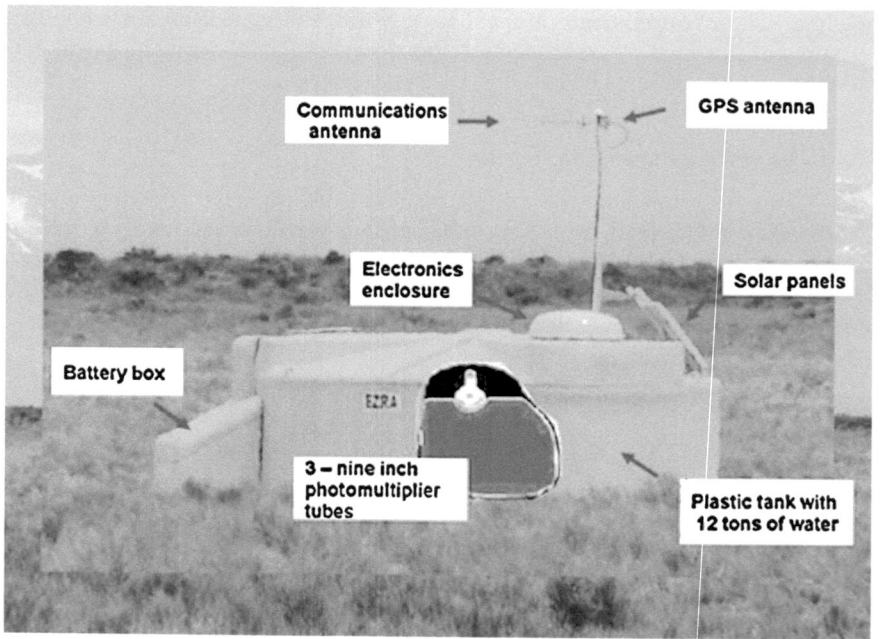

«Each one of these plastic tanks is filled with 12 tons of ultra-pure water. When charged particles from a cosmic ray shower zip through the tank, they emit tiny flashes of light which are seen by three very sensitive photomultiplier tubes. These tubes convert light into electrical signals. The source of power is the sun: solar panels are used to charge batteries, which provide all the power the tank needs. Data processing electronics, mounted inside a dome on top of the tank, collects the phototube signals and transmits the processed information via antenna to the main campus. A GPS device provides accurate timing, so that signals from many tanks can be properly compared».

The light that can be detected with the help of the photomultipliers is Cherenkov light. In the next two chapters we deal in detail with the detectors of Cherenkov light. For this reason, a more detailed description of the function mode of the detectors of the Pierre Auger Experiment is not needed. But since the Cherenkov light plays an important role in the experiments, to which we devote special attention in this study, its origin must already be explained briefly at this point: when a charged particle moves through a dielectric medium, the electron shells of surrounding atoms become polarized. If the particle moves relatively slowly in relation to the speed of light, no dipole field is produced, because the polarization of the medium around the particle is symmetrical. However, if the particle moves faster than the light in this medium, this symmetry is broken and a dipole field develops along the trajectory of the particle.

The dipole field generates electromagnetic waves that mainly overlap in a destructive way. In case of a forward direction of the particle, however, the superposition of the waves is constructive, and the result is a light cone with an opening angle α, which accompanies the particle in its trajectory[110]:

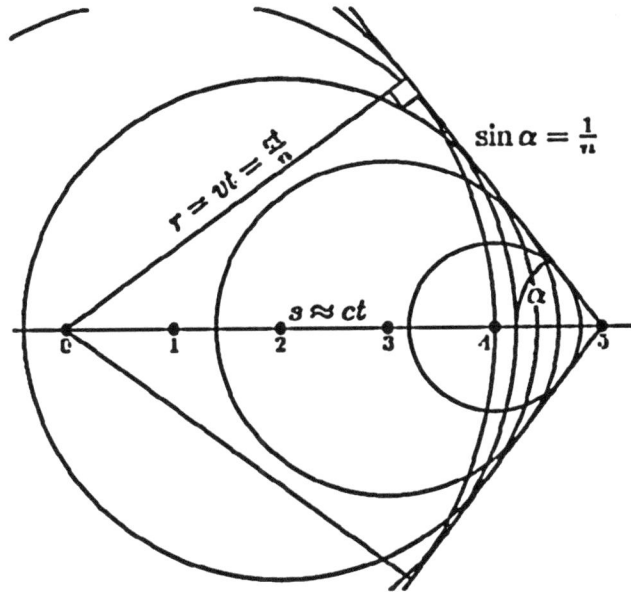

The Cherenkov - Light of the secondary particles of a particle shower that enters the water tank of a detector with higher speed than the light in that medium, can then be detected with the help of photomultipliers.

In order to exclude coincidental signals of the detectors (or signals not due to cosmic radiation), the information from the different detectors is filtered by a complicated five - trigger system:

> «Two local triggers are used: a simple "Threshold" (Th), and a "Time-over-Threshold" (ToT) which requires a lower but more extended signal (at least 12 FADC bins) and is more sensitive to the electromagnetic (EM) component

---

[110] Source: http://commons.wikimedia.org/wiki/File:Cherenkov2.svg

of the shower. A global trigger (T3) then asks for a relatively compact configuration of local triggers compatible in time with the arrival of a shower front. Finally, offline criteria are applied to reject accidental triggers ("physics trigger", T4) and to ensure the reconstructibility of the events ("quality trigger", T5).»[111]

The water tank detectors are in continuous operation. Based on the events filtered by the trigger system, one tries to reconstruct the spectrum of high-energy cosmic rays. For this purpose it is necessary to derive the energy of the primary particles, based on the ground events, by computer simulations. This step is not unproblematic, because the above-mentioned lack of knowledge of the dynamics of the particle showers. Therefore, one tries to achieve an independent (and nearly calorimetric) measurement of the energy of the particles with the help of the fluorescence detectors. This method is more direct and less problematic as the calculation of energy with the help of simulations. However, it has the disadvantage that the fluorescence detectors only work with weak moon light, and that the calibration of these detectors in view of the changing atmospheric conditions is not trivial.

By combining the fluorescence detectors and the Water-Tank-Cherenkov-Detectors, one tries to overcome the shortcomings of each of the detection methods.

3.5 The Image Generating Air-Cherenkov Telescopes

As mentioned before, the so-called «Cherenkov - light» emerges when a charged particle moves through a dielectric medium with a higher speed than the light in that particular medium. The gamma photons are not actually charged particles, but their entry into the Earth's atmosphere triggers particle showers that include electrons and positrons with suitable properties in order to produce Cherenkov - light.

Interestingly, the particle showers induced by gamma photons do not have large lateral expansion. This feature allows the reconstruction of the flight direction of the original photon of a shower, if the corresponding Cherenkov light is detected. In addition, it is possible to calculate the energy of the primary gamma photon based on the characteristics of the Cherenkov flashes.

The «Image Generating Air Cherenkov Telescopes» are thus built for the purpose of obtaining such information about the gamma induced Cherenkov flashes, because this way one can examine the gamma-radiation in the TeV range, an inaccessible range for satellite and balloon experiments (due to the already mentioned restrictions in the detector surface of such experiments).

---

[111] VAN ELEWYCK (2007).

The Cherenkov-telescopes consist of a mirror (or rather of a structure of hundreds of mirrors), a camera, which is installed in the focus of the mirror, a mobile stand or framework in which both the mirror and the camera are installed, and a number of electronic components, which transfer the signals from the camera into the computer, where the information is processed.

Here we can ignore the details of these elements as they are discussed in the next chapter using the MAGIC telescopes as an example. At this point, it therefore suffices to note, that the used devices (mirrors, camera components, electronic components etc.) must be able to detect extremely short and weak light flashes, because e.g. a gamma photon of 1 TeV energy produces only about hundred Cherenkov photons per square meter, and a gamma induced Cherenkov flash has an average duration of 1-5ns. Consequently, it is necessary to use large mirrors and very sensitive photodetectors in the camera. And it is equally necessary to provide the camera with a fast «data acquisition system». These requirements, however, are technically feasible.

However, the separation from signal and background becomes problematic, too .Because the hadrons of the charged cosmic radiation also cause Cherenkov flashes and the hadron-induced events in the Cherenkov telescopes stand in a ratio of about 1000/1 to the gamma induced events. In order to detect significant signals despite such an unfavourable signal-to-background ratio, it is necessary to perform a complicated analysis of the raw data. To what extent the theoretical expectations can affect this analysis; is a question to be discussed in the next chapter. However, it should be noted here that there currently are many parallel observations of gamma sources done by different research teams with different telescopes and with the help of a variety of analysis chains. At present there are four large active Cherenkov observatories for cosmic gamma rays: CANGAROO in Australia, VERITAS in the USA, HESS in Namibia and MAGIC in Spain. Further observatories are under construction or planned in India, China, etc. Under these circumstances and because the various groups are not only independent but stand in a fierce competition with each other, it seems not to be easy to imagine how the different data analysis procedures could all be influenced in a certain direction by theoretical trends.

3.6 The Underground Detectors

In this subsection, we will consider the techniques used for detecting cosmic neutrinos. In the last section it was already mentioned that the detection of these particles is an extremely difficult process due to the following two reasons: The first reason is, that the cross section of neutrinos is very small. The second reason is, that in the atmosphere of the Earth many particle events take place, which create a very strong background during the search for traces of cosmic neutrinos.

In order to avoid the problem of the cross section, it is necessary to build very large detectors. In order to minimize the atmospheric background caused by the charged cosmic rays, it is appropriate not to build the detectors on the Earth's surface but in the underground.

For this reason, the first neutrino detectors have been built inside mines. This, for example, was the case in the *SuperKamiokande* experiment in the Kamioka mine in Japan, where neutrinos from the sun and the supernova 1987 were detected. But to achieve larger detector volumes, efforts have been increasingly concentrating on experiments in water and ice in the last years. Experiments in water are for example *Baikal* (a Russian neutrino experiment in the Baikal See), *Antares*, *Nemo* and the planned *Nestor* experiment in the Mediterranean. The first detector built in the ice of Antarctica was *Amanda*. At present, a much larger detector is being built around *Amanda*: the *IceCube*.

All these experiments employ a similar technique, with the help of which the neutrinos can be indirectly identified through the detection of associated muons. A muon can emerge when one of the neutrinos passing through the detector interacts weakly with a nucleon. The resulting muon cause Cherenkov light in the detector and this light can be detected with the help of chains of photomultipliers.

The neutrino detectors, thus, consist of a large water or ice surface, on which a number of chains of photomultiplier [strings] are installed, which transmit the received signals optically and electrically to a data processing system on the surface of the detector. An essential component of this data processing system is a combination of different trigger systems that determines which events are finally saved. This structure can be seen in the following scheme of the *IceCube* detector[112]:

---

[112] Source: http://icecube.wisc.edu/ (reproduced with permission).

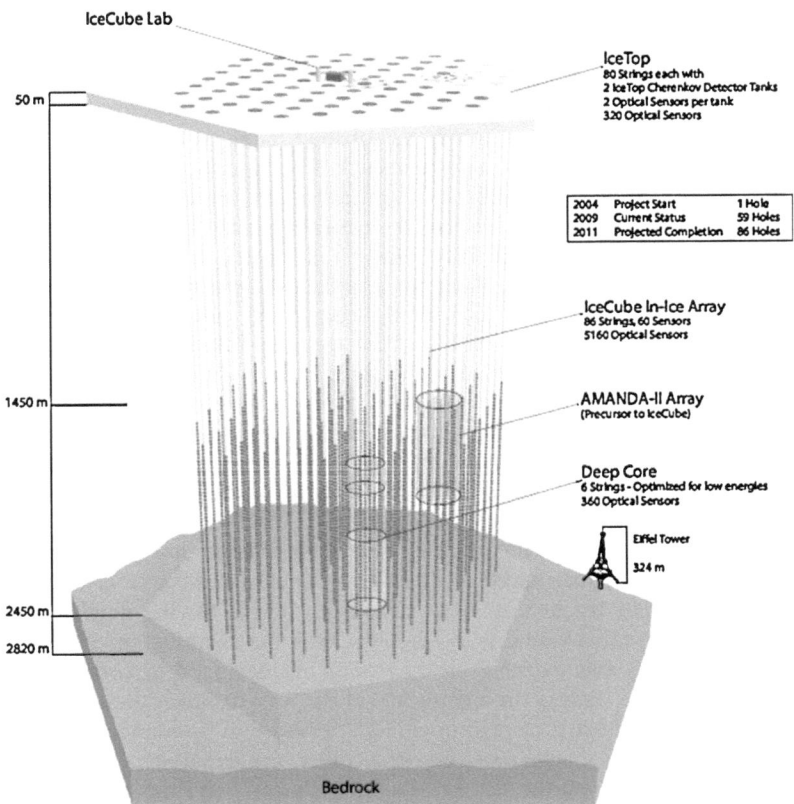

The problem of separating signal and background, which we have already dealt with during the consideration of numerous experiments of astroparticle physics, is a particularly serious one in neutrino experiments. The problem is so serious that, for now, there is no guarantee that neutrino signals from cosmic sources, other than the sun and some supernovae, can be successfully detected. A detailed analysis of the *IceCube* experiment can be found in the 4th chapter.

3.7 The Experiments of Astroparticle Physics and the Discussion between Realism and Social Constructivism

In order not to lose the thread of our philosophical inquiry, it is adviseable to ask at the end of this long and technical section, what useful information for the debate between realism and social constructivism can be won from considering

the astroparticle physics experiments. In order to be able to answer this question, we must again point out the different roles played by the experiments in the realistic and in the constructivist analysis of the dynamics of scientific research:
From a realistic perspective, the new phenomena of physics are discovered in the experiments with the help of already established methods of measurement and data analysis. The techniques that are used in the experiments were previously developed and checked regardless of the theories and models, which are to be tested by these experiments. Likewise, the data analysis is largely independent of the new trends in theoretical physics. These trends can at most provide possible interpretations of the examined data, but they cannot determine the results of the data analysis. Consequently, it can be argued that the phenomena are obtained largely «bottom up» –that is, starting from the raw data and with the help of well-established theoretical previous knowledge. So, nature reveals itself by the data, from which the phenomena are won. And the role of the theoreticians is limited to the development of models and theories that aim to explain the phenomena discovered in the experiments, starting from a hypothesis about the nature of physical reality (and/or about the structure of a given physical system).

From a constructivist perspective, however, the role of theories is more significant: the theoretical traditions generate tempting scenarios and mathematical ideas, which serve as a guide for experimentalists. It is not just that the experimentalists try to test these scenarios –a role for the new theoretical proposals that no realist would deny– but that both the planning and construction of the experiments and the analysis of results are designed to confirm certain theoretical notions... unless the nature leaves no way to reach the desired goal. But if a theoretical idea is appealing enough, a way will be found to confirm it. In other words, from a constructivist perspective, the dynamics of the research are largely controlled by the theory –i.e. «top-down»– and the phenomena are built so that they meet the preferred theoretical ideas. Naturally the case can arise, that an acceptable construction of the wished phenomena is not trivial, and that, therefore, the experimentalists must work hard, until a confirmation of the theory is reached, which is compatible with the standards of experimental physics. But the end result of the process is predictable: an interesting theory procures its own phenomena.

Regarding these two alternatives the analysis of the experiments of astroparticle physics shows the following:

Firstly, it seems justified to assert that a hypothetical control of the experimental results by theoretical trends –if such a control actually takes place– cannot be done by managing the characteristics of the experimental devices and the measuring instruments. Contrary to the opinion of some constructivists such as Collins or Pickering, awarding a certain plasticity to the detectors themselves

2. Chapter: Astroparticle Physics: An Overview for Philosophers 93

–in the sense of a certain adaptability of the instruments and the understanding of their function mode to the theoretical expectations[113]–, it must be said that at least in the field of astroparticle physics this plasticity is hardly or not at all present. Some reasons for this are as follows:

(1) The measurements in astroparticle physics are taken by devices whose fundamentals and function-mode are well known and constitute a safe technological basis. As we have seen in this section over and again, the basic equipment of these experiments are calorimeter, spectrometer, photomultipliers, mirror systems, etc. Their physical principles have long been investigated, and are used in different variations in many other experiments of physics and other sciences. Certainly, the planning of each new experiment makes certain technological developments necessary, and that means that the instruments used in the new experiments are usually not a mere copy of the old devices. But the fundamental ideas (e.g. about the function-mode of a calorimeter[114]) hardly change. In addition, the improvements of the instruments are always carried out on the basis of new, but already in other areas established techniques. (Regarding the used equipment it can be argued that the experimentalists are very conservative scientists). However, all this previous knowledge strongly limits the freedom to interpret, when a measuring instrument is working properly.

(2) In addition, all instruments are subjected to rigorous calibration procedures. This fact was mentioned in the description of some experiments outlined before, and we will consider this point closer in the next chapter (using the calibration procedures in the MAGIC experiment as an example). Certainly, the calibration possibilities in the field of astroparticle physics are limited, since it is not possible, for example, to generate particle beams with the energy of the most energetic particles of cosmic radiation in the laboratory. But, as far as possible, the measuring devices of the different experiments are calibrated again and again, which further limits the freedom of interpretation concerning the question of when a measuring instrument functions correctly.

(3) Finally, the new instruments are tested with regard to their ability to reproduce results that are already well known and are within their detection possibilities. Such results are the phenomena described in the previous section. A satellite

---

[113] Collins asserts for example that, apart from an agreement after negotiations between the specialists, there are no criteria for determining whether a new measuring device is working properly or not. And this means in essence that one can build devices that deliver the desired results depending on the case. [See COLLINS (1985), chapter 4]. Pickering for his part states that the instruments are flexible resources that can be adapted to some extent to fulfil the theoretical expectations. [See PICKERING (1987)].

[114] Zu diesem Punkt siehe z.B. WIGMANS, R. (2000), *Calorimetry* (Oxford 2000, Oxford Science Publications).

for the study of the gamma ray bursts must, for example, be able to detect gamma flashes, which are observed with other devices. A detector for the study of the charged cosmic rays must be able to confirm the well-known power laws of the spectrum of the cosmic rays, etc.

For clarity, i.e. to avoid constantly interrupting the presentation of the experiments with additional information about calibration procedures and other tests concerning the correct functioning of the instruments, the details of these tests are ignored here. (Only with regard to the satellite experiments, short quotations about the kind of checks with whose help the experimentalists explore whether the instruments are working properly or not were introduced). Nevertheless, in the next chapter I will outline some concrete examples of the tests which have been used for checking the operation of both the individual components of the MAGIC telescopes and the telescopes as a whole. These examples will confirm the inadequacy of the detectors as means to control the experiments in order to prove new theories.

But while the instruments themselves do not seem to be the appropriate means of a hypothetical control of the research process on the part of theoretical trends, the possibility cannot be ruled out at this point of our investigation, that the analysis of the raw data obtained in the experiments is influenced by certain theoretical expectations. Since we have seen in each example that the phenomena can only be obtained after a particularly difficult process of data analysis. A reason for this is, that in most experiments of astroparticle physics the signal-to-background ratio is very unfavourable. To find significant signals with a signal-to-background ratio in the order of $1/1000$ (or even worse) requires a very careful separation process which might eventually be disputable. Moreover, a complex process of data analysis is also required due to the fact that the variables measured in the experiments are only very indirectly connected with the variables the physicists need to describe and explain the various phenomena. One is interested e.g. in the energy and flight direction of the primary particles of the cosmic radiation, but what the detectors register are for example Cherenkov photons from the particle showers caused by these primary particles. Thus, do theoretical expectations affect the data analysis of the experiments of astroparticle physics? This question cannot be answered in a general way, but it is necessary to descend to the precise details of each analysis. Certainly such a complete review of the different methods of data analysis cannot be implemented here. But we are going to examine with more detail in the next two chapters the question of the possible theoretical control of the data analysis on the example of the experiments MAGIC and IceCube.

At this point, only the following general comments should be made: if one looks at the large number of experiments and observatories at present actively contributing to the advance of research in the field of astroparticle physics, it quickly becomes apparent that many comparable observations are made by different research teams, which work completely independent from each other. Consider, for example, the work of research teams in the area of large Cherenkov telescopes and Observatories (MAGIC, HESS, VERITAS, KANGAROO). Each of the research groups performs the data analysis with partially different methods completely independent from each other. Even within a scientific collaboration the various associated groups sometimes use different methods of analysis.

But when, in spite of this fragmentation of data analysis, the individual groups come to the same result, it seems difficult to understand, how theoretical expectations should influence the outcome of the analysis carried out by different groups independently from each other. In addition, the general recognition of a new phenomenon only happens when several independent collaborations confirm this phenomenon. And beyond that, each collaboration fears the loss of its reputation, if one of its research teams announces new phenomena, which cannot later be confirmed by other collaborations. Consequently, the results of the different analyses are very carefully revised within the collaborations before publication. Under these circumstances it is hardly conceivable how the theoretical trends can crucially affect the discovery of phenomena in astroparticle physics.

4. Theoretical Models in Astroparticle Physics

After the main phenomena studied in astroparticle physics as well as the diverse experiments and research methods of this field have been summarized, we now turn to the diverse theoretical models, which are connected any way with the phenomena of astroparticle physics.

In a first approximation we can divide these models into two groups: on the one hand, we are dealing with models, which try to explain the characteristics of the different kinds of sources of cosmic radiation, described in the 2nd section. (The theoretical models that attempt to explain the spectrum of charged cosmic rays as a whole –instead of the spectrum of a particular type of source– can be also counted among this group). On the other hand, there are models of theories or parts of physics, which are currently under construction, and may have some implications in the field of cosmic rays. This second group can, in turn, be divided into two components: firstly, there are models of dark (or exotic) matter, which e.g. imply effects in the composition of the cosmic radiation; and secondly, there are models of quantum gravity, which include some effects concerning the arrival times of gamma rays from the gamma-ray bursts, or from the flares of active galaxies.

Therefore, this section will be divided into four subsections: three of them are dedicated to the just-mentioned types of theoretical models, while the fourth summarizes some notes for our comparison between the realist and the constructivist view of the dynamics of scientific progress, which can be derived from the consideration of the theoretical models of astroparticle physics.

There is a further kind of theoretical models, which is not discussed in this section (for the sake of simplicity). These are the models that describe the details of the propagation of the particle showers in the atmosphere and the detection of the particles in the different experiments and telescopes. Some examples of these models, however, will be discussed in the next chapters.

4.1 Models of Sources and Spectra of Cosmic Radiation

The first group of theoretical models is motivated directly by the phenomena mentioned in the second section of this chapter. Therefore, the models of this group are not conceived to open a new field of fundamental research. They rather aim at explaining the phenomena of astroparticle physics with the help of the existing theoretical tools.

As noted in the second section, the phenomena of this field are (1) the composition of cosmic rays, (2) the form of the sources (point-like or extended), (3) the spectra of the sources (and of the cosmic radiation as whole) and (4) the possible variations of particle fluxes of the sources.

Thus, the theoretical approaches have to answer questions such as the following: how can the composition of the cosmic radiation be explained? How is it possible that the charged particles are accelerated in the sources in such a manner that the most energetic of them exceed the maximum energy of the particles in the largest particle accelerators of the world by several orders of magnitude? What is the cause of the periodicities and irregular flares of the sources? How can the power laws of the spectra be explained? What is the reason for the three different power laws of the spectrum of charged cosmic rays?

In order to answer these questions, different models are needed: one needs on the one hand, models of creation and acceleration of the cosmic particles that are not tied too closely to the specific characteristics of the sources, because many different types of sources exhibit certain similarities in the emitted radiation, possibly due to the effects of similar acceleration mechanisms. But, on the other hand, models are needed that take into account the specific structure of the sources. For many details of the phenomena cannot be explained by general mechanisms but with the help of additional assumptions about the nature of the various types of sources.

For this reason, I will at first present two examples of models that can be used to describe processes that occur in different sources, and then I shall refer to models that require specific considerations about the structure of the sources. Regarding the first group of models, we briefly consider the following:

(a) Models of the production mechanisms of gamma-radiation
(b) The model of shock acceleration and the explanation of the power laws of the spectra

(a) Models of the production mechanisms of gamma-radiation
The standard explanations of the gamma component of cosmic radiation is, that the gamma photons emitted from the sources originate from high energy particle interactions within these sources (or in their immediate environment). There are two different types of models for production of high-energy gamma-radiation: the hadronic and leptonic model.

According to the most common hadronic models, the gamma photons arise from the collision of high energy protons from the source with atomic nuclei of the interstellar medium in the vicinity of the sources. In inelastic interactions between high-energy protons and atomic nuclei, charged and neutral pions are created.

The neutral pions, however, are short-lived and decay into two gamma photons after about 10^{-16}s. At this point it is interesting to note that the hadronic mechanism implies a correlation between the gamma flux and the neutrino flux of the sources. The reason for this is that the charged pions, which arise in the same collisions as the neutral pions, decay quickly and neutrinos are a product of their decay. In case of a proton-proton collision, these processes are as follows:

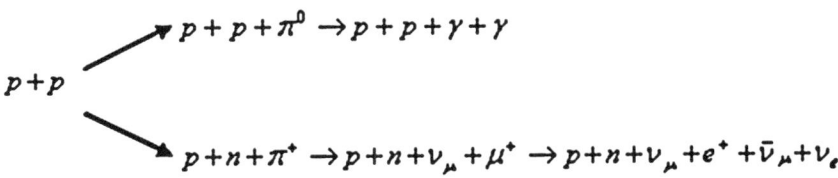

If some high-energy photons are present in the source, further gamma photons can be produced due to the interactions of these photons with high-energy protons. This process, in which an unstable delta resonance is first created, are as follows:

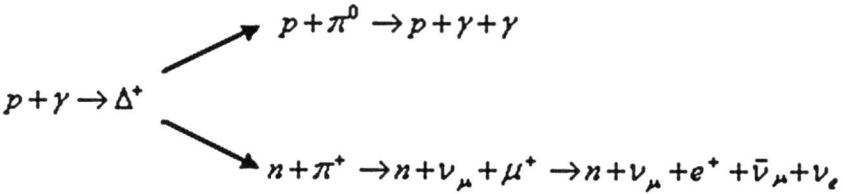

In this process there also is a connection between the production of gamma photons and the production of neutrinos. For that reason the evidence of an increased flux of neutrinos from a flaring gamma source could be seen as a clear indication of the effect of the hadronic acceleration mechanism... if only the required neutrino detectors were available.

The gamma photons could, however, arise from completely different processes, such as the *synchrotron radiation* that is emitted when relativistic charged particles move in magnetic fields, or *Brennstrahlung radiation* that is emitted when relativistic charged particles are deflected in the electric field of an atomic nucleus or an ion, or *inverse Compton scattering*, in which lower energy photons are scattered to higher energies by relativistic charged particles. The relativistic particles in all these processes could, in principle, be both hadrons as well as leptons, but due to the high mass of the hadrons extremely strong magnetic fields would be required for the synchrotron radiation, and no significant number of gamma photons could be produced by the other mechanisms, either. Therefore, it is assumed that the mentioned processes can be responsible for the production of gamma-radiation only in those cases where an electron population (or a positron population) is accelerated by a magnetic field.

The hypothesis that the gamma-rays are produced by synchrotron radiation of electrons plus inverse Compton scattering of photons, is called *Synchrotron Self-Compton Model* and is one of the most plausible explanations of the gamma emissions of many currently known sources.

The diverse processes that can lead to the creation of gamma photons can be represented by the following scheme[115]:

---

[115] Source: NADIA TONELLO, «Study of The VHE γ-Ray Emission from The Active Galactic Nucleus 1ES1959+650», Dissertation, Max-Planck-Institut für Physik München (2006) p.11.

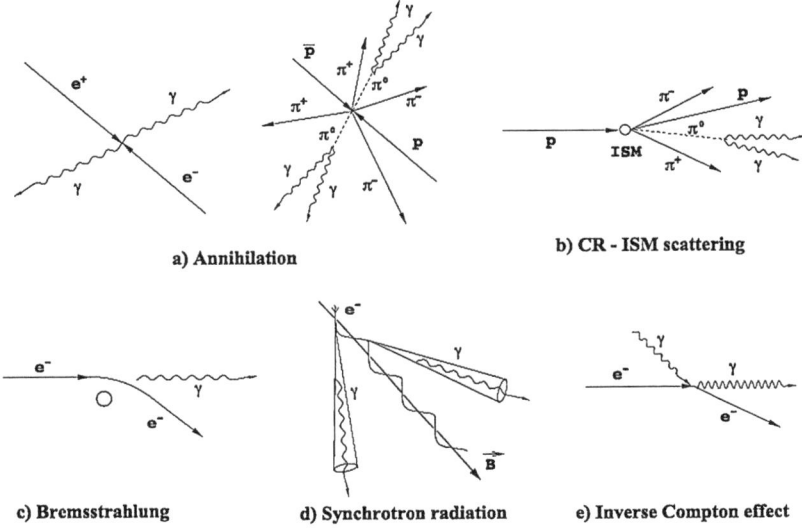

a) Annihilation b) CR - ISM scattering
c) Bremsstrahlung d) Synchrotron radiation e) Inverse Compton effect

I only excluded the annihilation processes, because they play no significant role in most models of sources of gamma-radiation.

(b) The model of shock acceleration and the explanation of the power laws of the spectra

The mechanism of the stochastic shock acceleration was already investigated in 1949 by E. Fermi. Nowadays it is believed that the stochastic shock acceleration is the key for the understanding of the power law behaviour of the energy spectrum of charged cosmic rays.

Fermi sketched a model for the energy increase and acceleration of a particle, which crosses a number of shock waves in plasmas. One distinguishes between the Fermi Acceleration of second order and the Fermi Acceleration of first order. In the Fermi acceleration of second order a particle with energy E_1 and impulse p_1 penetrates into a magnetized gas cloud and is accelerated, before it leaves the cloud with an energy E_2 and an impulse p_2:

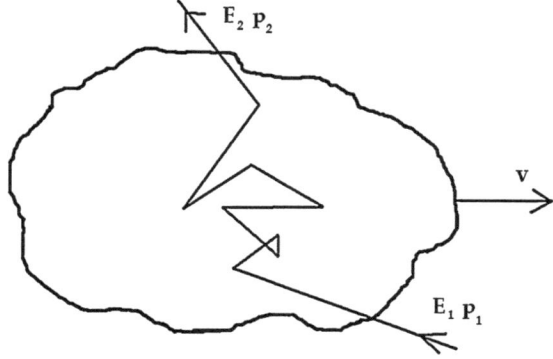

More interesting for astroparticle physics is, however, the case of the Fermi acceleration of first order. In this case a particle crosses not only some gas clouds, but a shock front, which move with speed V_s with respect to the environment:

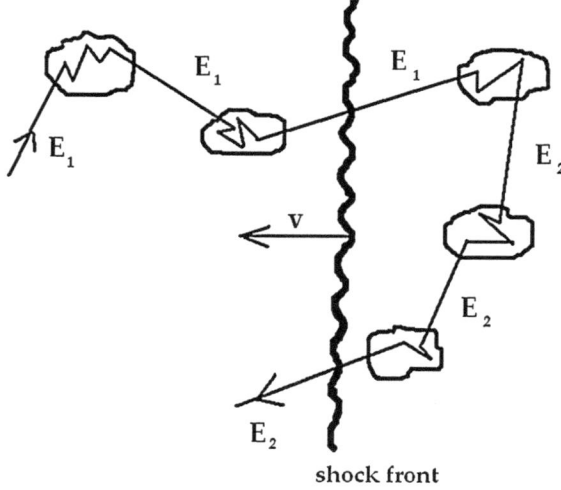

In the depicted movement, one assumes that the particle can go through the shock front *n* times and that its energy will increase ΔE each time. (The energy gain in the clouds outside the shock front is very small compared to the energy gain after crossing the shock front. For this reason, only the energy gain at the shock front is considered in the picture). One also assumes that the particle possesses a certain probability of escaping by each run of the front. This is the probability that the particle leaves the acceleration region after this run.

Under these assumptions, one can show that given a number *N* of particles, which are accelerated in this way, the number N_E of particles, which leave the region with larger energy than *E*, is as follows:

$$N_E \propto \frac{1}{P_{entk}} \left(\frac{E}{E_0} \right)^{-\gamma}$$

, where P_{entk} represents the probability of escaping, E_0 the initial energy of the particles and γ is a power indices. Consequently, the Fermi Acceleration Mechanism implies that the energy spectrum of the emited particles takes the form of a power law. At present the idea is, that this mechanism supplies the explanation for the power laws of the spectra of the cosmic radiation.

Let us now consider the theoretical models that include additional assumptions about the specific nature of the sources. It has already been mentioned at the end of the second section of this chapter that during the presentation of the phenomena associated with the different types of sources of gamma-radiation, hypothetical assumptions about the nature of these sources were combined with the description of the phenomena from time to time. Under the name «active galactic nuclei» there are, for example, a number of extra-galactic sources grouped, which exhibit actually very different characteristics. They were represented in this way because a plausible theoretical model (developed in the 90th) suggests that most of the differences are due to the sight angle from which we look at the different AGNs. According to this model an active galaxy simply is a galaxy, whose central, super massive black hole is currently in a phase of strong accretion. Perpendicular to the accretion disk two jets are formed, in which relativistic particles that cannot be incorporated in the black hole, are emitted.

This model appears plausible, and has the advantage of bringing together many different phenomena in a common framework. But it needs further theoretical work, as many details about the origin and acceleration of cosmic radiation in the AGN model remains unknown. Let us regard for example the application of the model to explain the characteristic double-peak structure of the spectra of blazars. The spectra of blazars have a low energy maximum in the radio to soft X-ray range and a high-energy maximum in the hard X-rays to gamma-ray range. Eichmann summarizes the explanation of these characteristics with the help of the current AGN of model as follows:

> «About the origin of the low-energy peak large agreement prevails. It is caused by synchrotron radiation of high-relativistic electrons and positrons. This radiation emerges in high-energy plasma clouds in the jets of AGN, which according to the kinetic two-phase plasma model [Schlickeiser (1996)], are emitted at certain times from the central region. In the quiet phase, in which the AGN is most of the time is, charged relativistic particles are accelerated in the potential of the black hole, while they leave the accretion-region outward along the rotation axis in a second, short and jerky phase. Thus, the intergalactic medium is enriched with high-energy plasma clouds of high pressure, the so-called blobs or plasmoids. These cross the comparatively thin intergalactic medium with high-relativistic speed and form the observed jet. [...].
>
> There are two different models [for] the composition of the plasmoids. On the one hand, the leptonic model, according to which the relativistic particles in the blob are high-energy electrons and positrons and, on the other hand, the hadronic model, according to which relativistic hadrons (essentially protons) are postulated as additional ingredients. [...] Two-stream instabilities, which are generated by the speed difference of the particle streams, produce aperiodic magnetic field fluctuations. The incident particles become isotropic and lead to an approximately monoenergetic Pick-up [Schlickeiser and Lerche (2007)]. In the resting system of the blobs the relativistic electrons and positrons rotate around the magnetic field lines and produce the low-energy synchrotron radiation responsible for the peak in the low-energy range. [...]
>
> For the emergence of the high-energy peak in the broadband spectra of the AGN there are different approaches: according to the leptonic model soft photons are scattered by high-relativistic electrons thereby increasing their energy (inverse Compton Scattering). [...] The acceleration of the electrons on high-relativistic energy would have to happen very fast in order to compensate the radiation losses. For that reason one does not apply electrons and positrons as primary particles in the hadronic model, but assumes that they are just produced by the inelastic collision of two primary hadrons –usually protons–»[116]

As we can see, this kind of explanation connects the above-mentioned general theoretical models about the emergence of gamma radiation with additional

---

[116] EICHMANN (2009) 6-8.

elements, which are based on the consideration of the specific structure of the AGNs. Such models have also been developed for the explanation of other characteristics of the AGNs and for the explanation of the characteristics of the different sources of cosmic radiation. They are usually produced with the help of established theoretical tools, and try to reproduce all details of the phenomena of the sources as precisely as possible.

4.2 Models of Dark (and/or Exotic) Matter

There is a second group of theoretical approaches, which can entail possible consequences for the field of astroparticle physics. These approaches are partly motivated by astrophysical observations and partly by theoretical considerations from the fields of cosmology and particle physics. They are the models of the so-called «dark matter» and the models of other exotic particles (like e.g. magnetic monopoles), which at present, however, get much less attention by specialists. Therefore, we mainly focus the models of dark matter.

The idea, that a large part of the matter of our universe is invisible, first emerged as a result of the observation of the rotation curves of spiral galaxies. For these galaxies rotate in the peripheral area much faster than calculated on the basis of Kepler's laws and the visible distribution of matter[117]. This phenomenon could be seen as an indication that the laws of gravity must be modified. But it can also be alternatively interpreted as indication for the existence of a halo of dark matter around the spirals. Additional evidence for the existence of dark matter comes from the field of cosmology. Measurements of the primordial abundances of light elements combined with detailed calculations of the nuclear synthesis in the early universe imply that the baryon density is $\Omega_b \approx 0.04$ [118]. Nevertheless, recent calculations of the matter density of the universe give a much greater value $\Omega_m \approx 0.3$.

Consequently, one can draw the conclusion that the matter in the universe is not «normal» baryonic matter for the most part. This result is also confirmed by computer simulations of structure formation in the cosmos. For the best match between the simulations and the observed large scale structure of the cosmos is achieved if one assumes that the formation of galaxies and galaxy clusters takes place in the walls of a cosmic lattice of dark matter. And finally, one finds evidence for the existence of kinds of particles, which have the assumed properties of dark matter, (in particular the property that they interact only

---

[117] See e.g. SCHNEIDER (2006) 100 ff.
[118] See e.g. SCHNEIDER (2006) 166-167.

weakly and gravitational) in the theoretical attempts for the extension (and/or overcoming) of the standard model of particle physics.

The attempts to build a bridge between the issue of dark matter and astroparticle physics reveal an interesting interplay between theoretical and experimental elements, because the aim of these attempts is not to examine whether an already fully elaborated theory of dark matter can be tested experimentally. It is rather to investigate whether there are phenomena in the field of cosmic radiation, which provide orientation in the jungle of theoretical possibilities (in the sense, that some of the many candidates for the role of dark matter can be ruled out, while other candidates become sharper outlined with the help of these phenomena). In this context the observations are to serve not so much as confirmation of theoretical models of dark matter, but rather as help for the formulation of a precise model.

Astroparticle physics may be helpful for these purposes only if the dark matter particles can interact in such a manner that any traces of such interactions are detected in the cosmic radiation. This would hardly be the case if the dark matter would become apparent only by its gravity. It is therefore not surprising that the models of dark matter that currently get more attention are those in which the particles of dark matter convert either by mutual annihilation or decay into typical cosmic rays. However, there is a profusion of such models[119].

The hypothetical particles postulated in these models are called WIMPs [weakly interactive massive particles]. Depending on the model, two, four or more symmetrical particles such as W^+ W^- or e^+ e^- arise after the annihilation or decay of WIMPs.

A hypothetical particle that got much attention recently, is the so-called neutralino χ. The neutralinos are the lightest stable particles in the supersymetric models [SUSY] of particle physics. In most models, the neutralinos have a mass in the order of some hundred GeV/c², and they can generate (by annihilation or decay) hadrons, leptons or gamma photons.

Typically, these processes generate, directly or indirectly, gamma-radiation. However, some production processes of this radiation, like e.g.

$$\chi + \chi \rightarrow \gamma + \gamma$$

or

$$\chi + \chi \rightarrow Z^0 + \gamma$$

are strongly suppressed in most scenarios, while other processes, such as the production of hadrons, are associated with the production of low energy γ-radiation, which is not easy to detect. Therefore, the theoretical models, in which the

---

[119] See e.g. MEADE et al. (2009), in particular p.5 and the references on this page.

following process plays a relevant role, have been getting special attention in the last years:

$$\chi + \chi \rightarrow l^+ + l^- + \gamma$$

Where l^+l^- represent leptons.

The reason for this is that the gamma photons generated this way should have an energy that lies in the detection range of the Cherenkov-telescopes. Thus, if the dark matter interact in such a way, it should be possible to detect (with the help of current Cherenkov-observatories) some gamma flux from the galactic halos –especially from the halos of dwarf elliptical galaxies, which are very small and dark.

However, the chances of this detection method depend on the size of the actual gamma flux, and there are very different calculations of this flux. In principle, it can be stated that the annihilation models imply a gamma flux proportional to the square of the dark matter density, while the decay models imply a gamma flux proportional to the dark matter density[120].

As no significant gamma flux from the halo of these galaxies has so far been detected, some theorists now suggest that the decay models are more plausible than the annihilation-models.

Besides the models for the generation of gamma-rays from neutralinos, there are other theoretical approaches that try to calculate the influence of the annihilation or decomposition of dark matter on the number of electrons and positrons in the charged cosmic radiation. A disadvantage of these approaches is, however, that other astrophysical mechanisms can imitate the excess of electrons and positrons caused by the interactions of the dark matter.

Beyond that, some theoretical approaches for the computation of the neutrino flux have been developed in recent years, which could be caused by the annihilation of WIMPs in the Sun[121]. These neutrinos would be produced from the decomposition of hadrons or bosons generated from the annihilation of WIMPs, such as e.g. according to the following process:

$$\chi + \chi \rightarrow W^+ + W^-$$

Finally, let us note that theoretical models of other exotic physical entities such as topological defects (magnetic monopoles, etc.) –that are postulated in the diverse versions of the GUT [Grand Unified Theory]– describe the production of neutrinos and other particles that, in principle, should be detectable with the help of the tools of astroparticle physics. The magnetic monopoles have, for example according to the current theoretical representations a magnetic charge and a mass in the order of $10^8 GeV - 10^{17} GeV$. Magnetic monopoles with a mass of

---

[120] See e.g. MEADE et al. (2009).
[121] See e.g. HALZEN, F. – HOOPER, D. (2009) and the references included on this paper.

about 10^{15} GeV accelerated to relativistic speeds in galactic and extragalactic magnetic fields would emit 8400 times more Cherenkov-Light than a relativistic muon. Such a phenomenon should be detectable with current Cherenkov-Detectors such as IceCube[122].

4.3 Models of Quantum Gravity

The third group of theoretical models that imply some consequences in the field of astroparticle physics are some models of quantum gravity.

The discovery of an empirically successful theory of quantum gravity can be described as one of the major dreams of theoretical physics. Because the discovery (or construction) of such a theory would imply that all physical forces are quantum-mechanically characterizable, and that the quantum theory represents consequently the universal framework for physics. With the establishment of a theory of quantum gravity, the completion of theoretical physics would be just a few steps away. And it would certainly be a very elegant completion.

Unfortunately, no complete quantum theory of the force of gravity has been formulated up to now (i.e. after several decades of intensive theoretical work), and it is not even guaranteed that such a theory can ever be established. Since, for example, no quantum-mechanical effects hat been observed in connection with the gravitation so far. That is, the classical description of gravitation, as it is accomplished in the general theory of relativity, seems to be a completely adequate characterisation of this force at this point. The hope is, however, that at the Planck Scale (i.e. at time intervals under 10^{-44} s and distances under 10^{-35} m) the gravitation will show their hidden, quantum-mechanical character, if someone succeeds to find out, how a trace of the events on this scale can be made «visible».

There are, however, different tentative quantizations of the force of gravity, i.e. different approximate models, which can be understood as simplifications, sketches or outlines of the main lines of the requested theory. What happens now, however, is that numerous models of quantum gravitation possess a common characteristic, i.e. they imply that the speed of light in vacuum is energy-dependent. In other words, they imply a violation of the so-called Lorentz Invariance. Usually it is a violation of this invariance in the sense of high-energy photons moving in the vacuum with lower speed than the low-energy photons. An approximate description of this effect (for the low energy segment) reads as follows:

---

[122] NIESSE, P. – SPIERING, C. – AMANDA-KOLLABORATION (2001).

$$c' = c(1 + \xi \frac{E}{E_p} + \zeta \frac{E^2}{E_p^2})$$

where ξ and ζ are model-dependent parameters, and $E_p = 1.22 \cdot 10^{19} \, GeV$ is the Planck Energy. One certainly assumes that the energy dependency of the speed of light, if it even exists, does not represent a large effect. Because, after all, the relativity theory, which includes the Lorentz invariance, describes all known phenomena of the light propagation extraordinarily well. But the distances to some gamma Ray Bursts and AGNs are so immense that, according to current computations, the arrival of the high-energy gamma photons from the flares of such objects should show a certain energy-dependent distortion..., if this effect is not obscured by other phenomena in the sources.

This circumstance opens the chance (particularly instructive for the philosophical observer) of an interaction between the theoretical traditions of the quantum gravity research and the experimentalist traditions of astroparticle physics. Because a proof of the violation of the Lorentz Invariance in the mentioned phenomena would imply that for the first time the theoretical work in quantum gravity gets some empirical support, which would give rise to hardly imaginable prospects for the further development of both the theory and some experimental research lines of astroparticle physics. Therefore, we have reached a point where the expectations of the constructivists and the realists concerning the future development of a research line will probably be different. But this point will be discussed in the next subsection.

4.4 *The Theoretical Models in Astroparticle Physics and the Debate between Realism and Constructivism*

To which extent are the theoretical models constrained by a facts-framework given by the experimentalists, and to which extent do the theoretical preliminary decisions determine the direction of the experimental results and even the results themselves?

In this section, we have ascertained that there are at least three different groups of theoretical models with different relations to the experiments in astroparticle physics: the models of sources and spectra of cosmic radiation, which are formulated with the help of well-established physics, the models the dark (and exotic) matter and the models of quantum gravity.

As to the first group, it is obvious that the theory is always falling behind the experiments. For it is the concrete empirical results that provide the starting point for the theoretical analysis. We will not find theoretical expectations, which should unavoidably be confirmed by appropriate experiments. Of course, each specialist has his favorite models, which he would like to confirm through the analysis of the data. But these models are not so significant that they are able to necessarily turn the development of experimental physics in a certain direction, amongst other things because there is no consensus about which models are to be preferred. The favorite models of some specialists are not the favorite ones of others. And this guarantees a close examination of each interpretation of the data.

Concerning the models of the dark matter, the situation is more complicated. Because, on the one hand, we already find certain theoretical expectations, which are addressed to the experimentalists: They should find evidence for the existence of particles in the cosmic radiation, which interact only weakly and gravitationally. But, on the other hand, the different theoretical models do not specify exactly which characteristics these particles are to have, and what effects they will cause in the field of astroparticle physics. It is not clear whether the dark matter consists e.g. of neutralinos or other hypothetical particles, whether these particles generate cosmic radiation by annihilation or decomposition, whether the final products of these processes are two, four or several particles, and of what kind they are, etc.

Under these circumstances is not evident, how a «top-down» construction of the phenomena in the sense of constructivism should take place. The viewer who examines the development of the research in this area rather gets the impression that what is really expected of the experimentalists is to bring some order into the confusion of the theoretical possibilities. And that means that the experimentalists enjoy a high degree of independence from theoretical trends. Surely every experimentalist can have his favorite models and theoretical ideas that he would like to confirm by means of his observations. But because in this case (as in the case of the models of the first group) no agreement prevails about the results to be achieved, it seems that a «bottom-up» development of the research (as suggested by the realists) almost required at first sight.

The situation would possibly be totally different, if some definite hypothesis of the dark matter would find considerable support from the theorists and if this hypothesis would have definite consequences in the field of astroparticle physics. In this case the suggestions of the theorists to the experimentalists would be explicit and clear. And then one would have to examine very carefully whether the further development of the experimental research on this topic is to be understood as a genuine discovery or rather as a construction of the needed phenomena.

But this is just the situation one finds when looking at the significance of the models of quantum gravity for the development of some lines of research in astroparticle physics. For the theoretical statements in this case are really explicit and clear: a proof of the violation of the Lorentz Invariance is to be supplied by the observation of flares of AGNs and gamma ray bursts. The importance of this result for the further development of the theoretical research program of quantum gravitation and for the possible development of a new experimental area for the specialists of astroparticle physics can hardly be exaggerated. For this reason it can be stated that the study of the present and future interaction between the theoretical research program of the quantum gravitation and the experimental searches of astroparticle physics represents an excellent field for the comparison and the evaluation of the realist and constructivist analyses of the dynamics of the research.

According to the logic of the constructivist approach, one would have to assume in such a case that astroparticle physics will, sooner rather than later, provide a confirmation of the expectations of the theorists. But from a realistic point of view –considering, among other things, the fact that many details of the emission processes in the sources are still unknown– will the task of proving quantum gravity still take (at best) a long time. Because as long as the emission mechanisms of the gamma photons in the gamma ray bursts and in the flares of the AGNs are not clarified, it would not be justified from a realistic point of view to regard the quantum gravitation as confirmed by data from such sources (even if some observations would point into the preferred direction... not to mention if it is not the case).

Realism or constructivism? It is still too early to decide which of the two analyses describes the development of the interaction between astroparticle physics and quantum gravity better. Nevertheless, there already are some indications that seem to favor the realistic analysis. But this is part of the next and last section of this chapter.

5. Final Comments on the Discovery and Construction of Phenomena in the Field of Astroparticle Physics

Are the phenomena of physics discovered or rather constructed? To a large part the debate between realism and social constructivism can be synthesized with the analysis of this question and the possible answers. If the phenomena are discovered, they ultimately correspond to nature's mode of being and, therefore, they tell us something about the physical reality. If the phenomena however are constructed, they only have to do with the preconceptions of scientists, who project rational constructions onto nature.

That the phenomena are discovered therefore is the thesis of realists, for which the efforts of the scientists serve as the gradual discovery of the details of this great (and not man-made) work of art, the world.

That the phenomena are constructed is in contrast to the thesis of the social constructivists, for which the efforts of the scientists are similar to the efforts of the artists who want to embody invented forms in a material body. And for this purpose they must learn, how the resistance of the material is to be overcome.

Thus, when Pickering e.g. speaks about the dynamics of research as a «dialectic of resistance and accommodation»[123], or when he admits that «how the material world is leaks into and infects our representations of it in a nontrivial and consequential fashion»[124], these statements are no concession to the realistic perspective. Because the influence of the material world are nothing but resistances of the researcher's «raw material» and the corresponding adjustments are nothing but resources, which the scientists-artists need to apply, in order to create the desired work of art.

Are the phenomena of physics thus discovered or rather constructed? In this chapter I have tried to give the basis for an answer to this question in the context of astroparticle physics. To this end firstly the most important phenomena examined at present in this research field have been here presented. Subsequently, the different detection methods and experiments used to study these phenomena have been briefly outlined. Finally, I have referred to the theoretical models that are being developed to explain these phenomena, as well as to other types of models, for which confirmation with the help of astroparticle physics is currently sought. Beyond that some notes concerning the connection between the different points and the topic of our study were added at the end of each section.

To conclude this chapter, the only task remaining is summarizing and completing these notes, to try to provide a first (provisional) answer to the question posed here. Let us briefly think about what the phenomena of astroparticle physics are: the phenomena characterized in this field are primarily spectra of the individual sources and of the entire cosmic radiation, as well as possible temporal variations of the particle fluxes of the sources, the form of these sources and the proportion of the different kinds of particles in the cosmic radiation.

Let us suppose that we want –probably unwittingly– construct one of these phenomena in a way that they support a particular theoretical model. In which places could we influence the experiments and observations in the desired direction? Prima facie it seems that we have two possibilities: we either influence the data acquisition by a clever construction of the measuring instruments, or we control the trigger-logic of the relevant detectors and the subsequent steps of data analysis in any way. Or we combine both procedures.

We have, however, noted in the third section of this chapter that arranging the material structure and function mode of the measuring equipment in favour of a theoretical proposal is (at least in the field of astroparticle physics) an extraordinarily difficult undertaking. For nowadays there is a profusion of practical knowledge about the calibration procedures and test methods of the

---

[123] PICKERING (1995) 22.
[124] Ibidem, 183.

instruments that are applied in this area, and the freedom of interpretation concerning the question of when a measuring instrument is working correctly, hence, is minimal[125].

Thus, it remains open the possibility of a theoretical guidance of the trigger systems and the whole data analysis. At this point further clarification is required. Because –as already noted in the third section– the data analysis in all experiments and observation types of astroparticle physics turns out to be very complicated, particularly due to the extremely unfavorable signal/background ratio.

Therefore, we must treat the question of the possible theoretical influence on the data analysis in astroparticle physics in a more detailed and separated way. We will turn to this task (using the example of the experiments MAGIC and IceCube) shortly.

Actually, the decision to postpone until the third chapter the discussion about data analysis (and the importance of data analysis to address the debate between realism and constructivism), is not free of problems. In fact, after reading the preliminary manuscript of this work, Prof. Allan Franklin suggested that this topic should be discussed earlier. For the problem of not doing so is that the reader might think that I hold the view that to have measuring devices operating properly (i.e. well understood and well calibrated devices) is enough to meet the constructivist challenge. But in reality this is not the case. The data obtained in experiments of this (and other) branches of physics have to be analyzed before conclusions can be drawn from them. And it is precisely concerning the data analysis where constructivists seem to be on safer ground. Then even "good data" can give incorrect results if the analysis procedures are incorrect.

So, although in the end I could not find a convincing way to advance this part of the discussion, I invite at this point the reader, not to extract from the previous presentation of the devices any precipitate judgment about the dispute between realists and constructivists in the field of astroparticle physics.

But let us provisionally assume that it would be possible to construct certain phenomena of astroparticle physics by means of the data analysis, following some previous theoretical choices. Which phenomena should now be constructed? It was noticed in the 4th section of the chapter that there currently is a particularly strong and well-defined theoretical pressure for a confirmation of the violation of the Lorentz Invariance in the flares of AGNs and gamma-ray bursts, and an equally strong, but not specifically model-based pressure for a confirmation of the existence of dark matter. Can we thus recognize a certain tendency to search for a confirmation of these ideas at all costs in the current evolution of the astroparticle physics?

---

[125] We have in the third section also noted that some concrete examples of these test methods will be outlined in the next chapter.

In my opinion the answer is NO. But we must immediately add that the processes are still open.

However, as an observer of these processes one rather gets the impression that the experimentalists act very carefully and cautiously with regard to the revolutionary theoretical hypotheses. That is, instead of trying to find an immediate confirmation, they rather try to impose constraints on the magnitude of the predicted new effects.

What they prefer (most frequently) is to try to show that the predicted new effects cannot occur in the order of magnitude as theoretically expected. And if some experimental results can be interpreted by a new revolutionary hypothesis, the experimentalists first try to examine and support more conventional interpretations. Two examples should clarify the typical attitude of the experimentalists at this point:

Example 1: *Current reports about constraints of the violation of the Lorentz Invariance*
The first theoretical models of quantum gravity, including a violation of the Lorentz Invariance, which could be confirmed by observations in the field of astroparticle physics, were formulated in the late nineties[126]. Meanwhile, it turned out that almost all the main approaches towards quantum gravity contain such a violation[127]. Nevertheless, this vigorous activity of the theorists did not meet with much response among the experimentalists. The collaborations of the large gamma-ray observatories (HESS, MAGIC, VERITAS, KANGAROO), only led to the publication of less than half a dozen articles dedicated to the analysis of possible of Lorentz Invariance violation on the basis of the flares of AGNs. This apparent inaction is not necessarily to be interpreted as a lack of interest, but rather as a sign of caution on the part of experimentalists. For they have no interest in presenting hasty results which may then be ruled out by further observations.

First notes concerning the violation of the Lorentz Invariance were presented in 2008 by the MAGIC collaboration[128]. They were based on the analysis of a flare of the AGN Markarian 501 in 2005. This analysis increased the confidence in the existence of the desired violation. For it suggests a possible energy dependence on the arrival of the gamma photons:

«The probability of the zero-delay assumption relative to the one obtained with the ECF estimator is P= 0.026. The observed energy-dependent delay

---

[126] See e.g. AMELINO-CAMELIA et al. (1998) 763.
[127] Shao, Xiao and Ma quote e.g. the following approaches: «spacetime foam, loop gravity, torsion in general gravity, vacuum condensate of antisymmetric tensor fields in string theory, and the so called double special relativity». SHAO - XIAO - MA (2009).
[128] ALBERT et al. (2008b) 253.

thus is a likely observation, but does not constitute a statistically firm discovery.»[129]

The authors of this paper, however, pointed out that the energy dependency on the arrival of the gamma photons does not necessarily imply a violation of the Lorentz Invariance. First, one should exclude that the cause of this result is a process in the source itself:

«We cannot exclude, however, the possibility that the delay we find, which is significant beyond the 95% C.L., is due to some energy-dependent effect at the source.»[130]

But all in all, the results of this analysis clearly pointed in the direction preferred by the theorists of quantum gravity. If there had been confirmations ob other collaborations at that time, one could have said that this development is not substantially different, than had to be expected from the constructivist perspective. But it went differently. A few months later the analysis of a flare of the active galaxy PKS 2155-304 in 2006 was published by the HESS collaboration[131]. And this analysis did not reveal any energy dependence on the arrival of the gamma photons. The contrast with the theoretical expectations could clearly be gathered from the abstract:

«In the past few decades, several models have predicted an energy-dependence of the speed of light in the context of quantum gravity. For cosmological sources such as active galaxies, this minuscule effect can add up to measurable photon-energy dependent time lags. In this paper a search for such time lags during the H.E.S.S. observations of the exceptional very high energy flare of the active galaxy PKS 2155-304 on 28 July in 2006 is presented. Since no significant time lag is found, lower limits on the energy scale of speed of light modifications are derived.»[132]

The derived constrains for a possible violation of the Lorentz Invariance were very strict, and they stood in direct contradiction to the results of Albert et al.:

«For a final verdict on this question further VHE observations of active galaxies are needed. However, the result already shows that the time delay reported for Mkn 501 in [Albert et al.], if considered significant, cannot be attributed to speed of light modifications. Current and future instruments such as Fermi for gamma ray bursts, or the proposed Cherenkov Telescope Array for active galaxies, will further improve the sensitivity of time-of-flight

---

[129] Ibidem.
[130] Ibidem.
[131] AHARONIAN et al. (2008).
[132] Ibidem

measurements, perhaps one day revealing deviations from Einstein's postulate.»

Now it was the turn of the FERMI collaboration, which could provide very accurate results from the observations of short gamma-ray bursts. Fermi was able to detect very energetic gamma photons from a short gamma-ray burst with measured redshift for the first time in 2009 [133]: GRB 090510. The detection of a 31 GeV photon in the first second of the flash has led the authors of the FERMI collaboration to put (at least temporarily) extremely narrow limits to the violation of Lorentz Invariance. They even claim that a whole class of theoretical models has been refuted by these constraints.

Therefore, the development in this line of research remains open. But the episode just summarized can be seen as an indication for the experimentalists in astroparticle physics not showing any special tendency towards constructing phenomena desired by the theorists. The next example can confirm this impression.

Example 2: *The surplus of electrons and positrons in the cosmic radiation and the dark matter*
In 2008 it was reported by the ATIC collaboration that by measuring the electron-spectrum of the cosmic radiation they found a surplus of electrons and positrons in the range of 300-800 GeV; a significantly higher number than theoretically expected[134]. Since then this characteristic of the electron-spectrum has been called «ATIC peak». The report of the discovery triggered a flood of publications. In July 2009, e.g., one could find over 200 papers in the ADS Databases in which the original paper of the ATIC collaboration was quoted. Most of these articles supplied interpretations of the described effect, and many of them pointed to the annihilation or decay of dark matter as a possible cause for the excess of positrons and electrons in the spectrum of the cosmic radiation.

Soon after that, the PAMELA collaboration reported that they have detected an increase in the positron fraction in the spectrum of cosmic rays in the energy range of 10-100 GeV. The report of PAMELA was in conformity with the report of ATIC and in direct contrast to the previously expected exponential decrease of the positron spectrum in this area. How could this new phenomenon be explained? Some authors suggested that the measurements of the positron fraction in cosmic rays provide an exaggerated result because it cannot be excluded that a certain number of protons are taken as positrons[135]. Most specialists assumed, however, that some unexpected source of electrons and positrons should be found. The dark matter could provide this source.

---

[133] ABDO et al. (2009b).
[134] CHANG et al. (2008) 362-365.
[135] See, e.g., SCHUBNELL (2009).

But interestingly enough, most authors emphasized the existence of conventional alternatives, which (at least to this stage) should be considered. Most people contemplated the possibility that the positron-excess is generated in nearby pulsars. However, other alternatives were discussed too, such as the production of positrons in the polar cap of exploding Wolf Rayet stars and red giants[136]. That is, a premature assumption of the dark matter scenario as «the» explanation of the phenomenon described has not taken place... although the theorists have a strong interest in confirmating the existence of dark matter. But rather than engaging in a rapid establishment of this hypothesis, the experimentalists tried to develop and examine all possible interpretations of the new phenomenon –in the spirit of what Franklin called «Sherlock Holmes strategy»[137]–.

The development of the discussion about possible explanations of the «ATIC peak» has become even more interesting in recent months, following the publication of new measurements of the positron spectrum of the HESS - Collaboration [in the range between 300 and 800 GeV] and the Fermi - Collaboration [in the area between 20 GeV and 1 TeV] because the new results suggest (1) that the surplus of positrons, if real, is not so prominent, as expected due to the results of ATIC, and (2) that this surplus of positrons is hardly compatible with currently favored annihilation-models of dark matter. In particular the Kaluza-Klein particles previously preferred by many theorists are not compatible with the results of HESS:

> «Despite superior statistics, the HESS data does not rule out the existence of the ATIC-reported excess owing to a possible energy scale shift inherent to the presented measurement. However there is no indication of a sharp drop in the cosmic-ray electron spectrum as expected in a KK dark matter scenario; the spectrum rather experiences a steepening towards higher energies and is therefore compatible with conventional electron populations of astrophysical origin within the uncertainties related to the injection spectra and propagation effects.»[138].

In the final discussion about the results presented by FERMI, the hypothesis of the dark matter is not seen as absolutely refuted –mainly because there are so many variations of this hypothesis. But the conventional explanation of the positron excess as originated from surrounding pulsar is clearly favored:

> «We found that the LAT spectrum can be nicely fit by adding an additional component of primary electrons and positrons [...]. Such an additional component also provides a natural explanation of the steepening of the spectrum

---

[136] BIERMANN et al. (2009).
[137] FRANKLIN (2009).
[138] AHARONIAN et al. (2009).

above 1 TeV indicated by HESS data. [...] [P]ulsars are the most natural candidates for such sources. Other astrophysical interpretations or dark matter scenarios cannot be excluded at the present stage.»[139]

Enthusiasm for the hypothesis of dark matter is not exactly what these sentences convey.

Conclusion: What can we learn from the examples mentioned above (and similar examples) is that the study of phenomena in the field of astroparticle physics is performed very independently of the favourite conceptions of the theorists. Of course, the developments described have not yet been completed. The exploration of these issues is ongoing, and it is not excluded that the dark matter or the violation of the Lorentz Invariance will eventually be based on phenomena of this research area. But the objections to the theoretical ideas that have been collected by the experimentalists so far, and the constant search for conventional alternatives to the revolutionary explanations suggest the conclusion that the process of testing the hypotheses of quantum gravity and dark matter, cannot be interpreted as a process of construction of certain desired phenomena... *at least not at present.*

«Discovery» and not «construction» seem to be the appropriate term for describing the derivation of new phenomena. And if one e.g. considers that the results of each research group are examined from other groups with similar or very different instruments and analysis, this conclusion could not be surprising. For the chances for a construction of the phenomena can only be (at most) very limited, because the teams know that the results that are achieved by a risky analysis of data, can be questioned by the other groups.

But thus we are again confronted with the topic «data analysis». And finally, we are faced with the question of whether (and to what extent) the data analysis in the field of astroparticle physics can be used to construct phenomena that are adapted to the new preferred theories. We will examine this issue in the next two chapters.

---

[139] ABDO, A. et al (2009c).

3. Chapter: MAGIC

1. Introduction

From the description of the most important phenomena, experiments and models of astroparticle physics in the previous chapter, we have gained some impulses for our examination of the realistic and the constructivist account of the dynamics of scientific research. In describing the various aspects of astroparticle physics, we have, however, encountered some issues that require more thorough examination than it is possible in a general overview of the subject.

It was, for example, pointed out that the observations in this field are performed by instruments which consist of well-known components that are used in various forms in many other contexts. And it has been claimed that the experimental physicist appears to be a very conservative scientist who always makes the improvements of instruments based on new but already proven techniques which have been implemented in other areas. Moreover, in this context it was pointed out that the meanwhile acquired practical knowledge about the calibration procedures and test methods of such instruments largely limits the freedom of interpretation concerning the question, when a measuring instrument is working correctly.

As evidence of these claims both were used in the previous chapter, the numerous and remarkable similarities in the structure of the different detectors – which consist of calorimeters, spectrometers, mirror systems, etc. that also apply in other areas of physics and their physical principles have already been explored– as well as some quotes from reports of PAMELA and Fermi-LAT which were mentioned as an example of the (reasonable!) approach to the development and calibration of the various instruments in this research field.

In my opinion, this information is already sufficient to draft an argument for the thesis of the unsuitability of the instruments as means for a theory-driven steering of the phenomena in the field of astroparticle physics. But without additional support, serious criticism can be raised against this argument which we will discuss in detail in the third section. To counter this criticism, it would be desirable to add concrete examples of how it is determined, whether these instruments function correctly during their development and calibration. Certainly such examples cannot be stated for each instrument mentioned in this study. For such a method could be applied only in the context of a much larger work. However, there is no doubt that the suggested argument against the theoretical control of the outcome of the instruments will seem more plausible, if the appropriate technical details are described more exactly at least in one case.

Another issue that should be examined more closely is the question of data analysis. In the previous chapter, we have mentioned that the separation between signal and background is a serious problem in all experiments of astroparticle physics. Because the ratio between signal and background is typically 1/1000 or even worse, implying that the phenomena in this area can only be found after a complicated process of data analysis. Under these circumstances, it is justified to ask the question to what extent such phenomena are discovered rather than constructed following certain theoretical preconceptions. A sound answer to this question can only be given, however, after a detailed review of the various methods of data analysis in astroparticle physics, a work that (also due to the lack of space) it cannot fully be realized here. Hence, it is meaningful to deal with the details of the data analysis at least in one case.

The goal of this and the next chapter, thus, is to describe in detail the instruments and data analysis in the case of two specific research collaborations in the field of astroparticle physics, and to derive the possible consequences of the considered aspects for the debate between realism and constructivism. This chapter deals with the MAGIC Telescope and the work of the MAGIC Collaboration, and the next chapter focusses on the IceCube Experiment.

To achieve the goal of this chapter, it is divided into the following sections: First [2. section] the structure of the MAGIC Collaboration is outlined. Immediately afterwards, [3. section] the MAGIC Telescopes are presented, and some details of the technical development of the telescopes are also mentioned. Then [4. section] follows a description of the two types of data analysis used by the members of the MAGIC Collaboration. And finally [5. section] some evidence for our discussion of the controversy between realism and constructivism are to be extracted from the discussion in the previous sections.

2. The MAGIC Collaboration

Let us begin with a summary presentation of the association of scientists, research teams and scientific institutions using the MAGIC Telescope. In December 2009, the MAGIC Collaboration consisted of about 217 scientists belonging to 20 research teams from Bulgaria, Germany, Finland, Italy, Croatia, Poland, Switzerland and Spain. Some teams are supported by universities (as, for example, *Technische Universität Dortmund, Universidad Complutense de Madrid, University of Siena, Universität Würzburg*, etc.) and some are located in other research centers (as, for example, *Instituto de Astrofísica de Canarias, Max Planck Institut für Physik* in Munich, *Institute for Particle Physics* at ETH Zurich, etc.). The aim of the collaboration is the development and use of the MAGIC Telescope which will be described in the next section.

Within the collaboration, the research teams operate to a large extent independently from each others. However, there are a number of committees, regulations and common activities that provide both, a more organized use of the telescopes and a coordinated development of the techniques and the different research lines.

The largest decision-making committee is the *Collaboration Board* which includes a representative from each research institute. A smaller group forms the *Executive Board*, responsible for the ordinary management and coordination functions. Other committees take care of specific tasks such as the distribution of observation time, the decision on joint publications of the collaboration, etc.

In addition to the committees, there are also thematic *working groups*. Some of these groups deal with issues that concern the development of data acquisition systems and data analysis, while other groups are occupied with the study of the sources of γ-radiation, etc.

Some people are responsible for particular tasks. These include, for example the *speaker* and the *coordinators* and *deputy coordinators* for physical and technical issues, for the applied software, for the maintenance of the telescopes, for the coordination with the IAC [*Instituto de Astrofísica de Canarias*], etc.

The different committees, the working groups and the persons with special tasks provide for a constant exchange between the research teams. Additional information exchange between members of the collaboration takes place mainly during meetings where the researchers of the different teams have the opportunity to make personal contacts and participate in lectures and discussions on various topics. Every year there are two major «collaboration meetings» which take place over a period of (approximately) four days, with all members of the collaboration being invited. The «collaboration meetings» are organized alternately by the different institutions participating in the collaboration. As part of this meeting, a conference of the *Collaboration Board* also takes place. In addition, there are specific meetings and workshops at irregular intervals, which are dedicated to a particular issue (such as the camera design for MAGIC-II, the development of new software for data analysis, calibration procedures, etc.). Finally, there are also seminars for the training of new members of the collaboration (or the training of any interested member): the *MAGIC-Schools*.

In view of such a structure one could ask (with view to the subject of this investigation), to what extent the existence of such large research partnerships can lead to a kind of community of interests of experimental physicists whose activities are not primarily determined by the will to discover the real features of nature. Would it not be more reasonable to assume that the dynamics of research collaborations is mainly determined by sociological factors, such as the necessity to justify the enormous costs of the equipment with the help of a particular discovery or the possibility of acquiring additional funding when interesting results can be presented? And should this be the case, this does not mean that large collaborations are particularly prone to the control of the research results by theoretical expectations? Because usually theoretical expectations determine, which results are noticed as particularly interesting and/or as innovative.

These questions and doubts are certainly justified. But I think they can not be answered by a general reflection about the social interests which are inseparably connected with the research work of collaborations. It is true that the volume of the funds used by collaborations is an indicator for a certain amount of pressure to succeed. And it also is true that such factors as, for example, the tendency to support earlier works of other collaboration members, could lead to a research dynamics in the sense of constructivism. But we must also take into account other factors that rather favour a realistic dynamics of research. These factors first include the fact –which has already been mentioned in the previous chapter–, that there are different (and competing!) collaborations which meticulously control each other. The publication of results which cannot subsequently be confirmed by other collaborations, can thus lead to the much feared loss of image.

Another factor that favours realistic dynamics in the research of collaborations such as MAGIC is the fact that the different research teams within the collaborations act very independently. In the 4th section of the chapter we will e.g. see that diverse teams have developed two different methods of analysis for the evaluation of the MAGIC raw data. Differences of this kind have the consequence, that also within the MAGIC Collaboration a certain competition between the teams and the scientists takes place, leading to mutual control.

Consequently, there are sociological factors which could support a tendency of the members of large collaborations to construct phenomena in the investigated area, and there are other sociological factors that favour a realistic attitude. What influences prevail, then?

It does not seem possible to answer this question a priori. It is, therefore, necessary for the purposes of this study to consider the approach of the MAGIC Collaboration in detail before we decide whether this work is to be interpreted as a discovery or as a construction of phenomena.

3. The MAGIC Telescopes

The aim of this section is to provide a description of the MAGIC Telescopes which have to serve as a basis for the discussion about the suitability of the instruments as resources for a theoretical control of the phenomena in the field of astroparticle physics. To this end, the section is divided into two parts: In the first subsection the main components of the telescopes are presented, while in the second subsection some concrete examples of the tests are discussed, which were performed in order to ensure the correct operation of both the individual parts of the telescopes and the telescopes as a whole. Some philosophical considerations are added at the end of each subsection.

[Before reading this section it is recommended to take a quick look at the points 3.4.1 and 3.5 of the previous chapter, in which the origins of the

Cherenkov light and the connection of this light with the particle showers produced by the gamma photon are discussed. Because this information forms the basis for the understanding of both, the structure and the goals of the MAGIC telescopes].

3.1 Components of the MAGIC telescopes

The MAGIC telescopes are two «Imaging Air Cherenkov Telescopes» located on the *Roque de los Muchachos* –at an altitude of 2220 m above sea level– on the Canary island La Palma[140]:

The name MAGIC is an abbreviation for «Major Atmospheric Gamma-ray Imaging Cherenkov telescope». The first telescope –MAGIC I– was inaugurated in fall 2003 and started to provide data in 2004, after a commissioning period of several months. The second telescope, MAGIC II, was inaugurated in April 2009 and is still in the testing phase. For that reason only the data analysis chain of the MAGIC-I is described in the next section.

The main components of the MAGIC Telescope are the following:

---

[140] Credit of the picture: Robert Wagner

(1) framework and drive system
(2) mirror
(3) camera
(4) calibration system
(5) «counting house»

The «counting house» is not really a part of the telescopes, but it is placed 162 m away from MAGIC-I. It must nevertheless be mentioned because essential components of the MAGIC Data Acquisition System have been placed there to minimize the weight of the telescopes.

A brief description of these four elements is as follows:

(1) The Framework

The dimensions and the framework structure of the two telescopes are identical. Each telescope has a height of 30 m and a weight of 60 tons. The frames of the MAGIC telescopes consist of very light carbon fibre rods. The use of this material has the effect that the telescopes can relatively easy and simple be moved despite their huge size. For this reason, it is possible to adjust the MAGIC Telescopes into any desired direction within 40 seconds. This allows MAGIC to respond quickly when a satellite reports the discovery of a gamma-ray burst[141]:

---

[141] Credit of the picture: Max Planck Institut für Physik, München

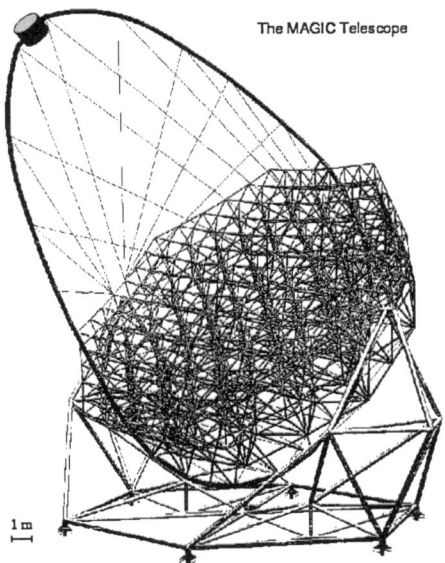

The MAGIC Telescope

(2) The Mirror

The mirror of each telescope is parabolic, because this form enables the reflected photons from the particles showers to reach the camera in the same order as they reach the mirror. Thus, the time structure of air showers remains and simplifies the data analysis. The mirrors of the MAGIC telescopes consist of a number of diamond-polished aluminium mirrors. MAGIC I consists of 964 single mirrors. They have a square shape of 49.5 cm edge length and their average reflectivity in the wavelength relevant for Cherenkov light (300-650 nm) is about 85%. The total mirror surface is 236 m^2, and the mirror diameter is 17 m. These data are similar for MAGIC II, although it should be noted that in the mirror of MAGIC II, some technical improvements were added[142]. So, the MAGIC telescopes are currently the largest Cherenkov telescopes in the world.

Regarding the dimensions of the mirror, it should be noted that the number of Cherenkov light Photons that are produced as companions of a gamma particle

---

[142] The biggest difference between the mirrors of the MAGIC I and II is that the new mirror elements have an area of 100 × 100 cm instead of 49,5× 49.5 cm.

shower, is strongly depending on the energy of the primary gamma photon, causing the particle shower. Therefore, the low-energy gamma photons can be detected only by large-dimensioned Cherenkov telescopes.

Thus, the following problem arises: the satellite detectors of γ-radiation can so far detect only photons up to energies of 10 GeV. But the usual Cherenkov telescopes cannot observe the energy range below 100 GeV. Consequently, the energy spectra of the sources of γ-radiation can only be partially investigated to date only in incomplete form. One of the main objectives of building the MAGIC Telescopes, hence, was to close that gap between satellite observations and Cherenkov telescope observations as good as possible. Currently, the MAGIC telescopes perform observations in the energy range from 30 GeV to 30 TeV[143]:

(3) The camera

The largest differences between the two MAGIC telescopes can be found in the design of the cameras. The MAGIC-I camera has a hexagonal area of 1,05 m diameter. This surface consists of 578 hexagonally arranged photomultipliers [*Photo Multiplier Tubes* or PMT] which convert the incident light into electrical impulses. In order to minimize the gaps between the PMTs (and thus the loss of

---

[143] Credit of the picture: Robert Wagner

light) a plate with reflecting hexagonal light funnels (the so-called *Winston Cones*) is installed above the PMTs. Each PMT together with its *Winston Cone* is called a pixel.

The camera of the MAGIC I consists of two kinds of pixels. On the inside of the camera, there are 396 pixels of 1 inch diameter (about 2,54 cm). They are surrounded by 180 pixels of 1.5 inch diameter, arranged in four concentric rings. This arrangement can be represented as follows:

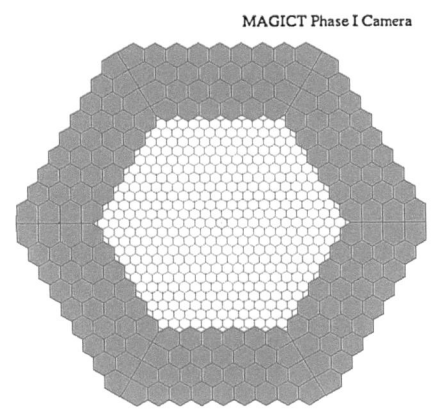

MAGICT Phase I Camera

The MAGIC II camera is designed based on the same general principle of MAGIC I. It consists of pixels, composed of PMTs plus *Winston Cones*. However, this is a new camera with 1039 pixels of 1 inch diameter, which are arranged uniformly. This arrangement can be represented as follows[144]:

---

[144] Credit of the picture: MAGIC Collaboration.

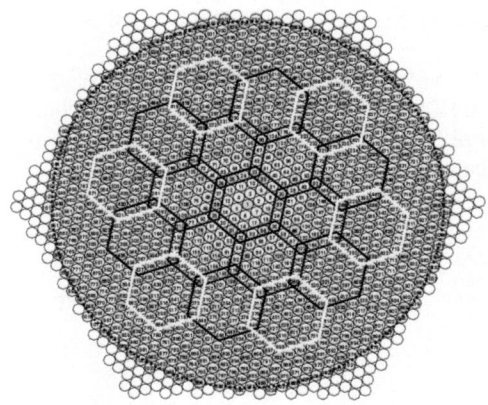

Currently, new types of photomultiplier tubes are tested. The use of *hybrid photon detectors* (HPDs) has been successfully tested, so that they could replace the usual PMTs in the near future. In addition, silicon PMTs are also under development. In the long run, the employment of these photomultipliers could increase the efficiency of the camera significantly.

(4) The Calibration System
In order to make reliable measurements of the Cherenkov light generated in the different atmospheric gamma events, the exact response of the photomultipliers from the camera must be known at any time. For this purpose a calibration system of the camera has been installed in the telescopes.
The system[145] consists of a number of *light emitting diodes* (LEDs) in three different colours, which send light pulses of different intensity to each PMT of the camera for calibration. The response of the PMTs is then stored in the «counting house» as calibration information, together with the raw data of the observations[146]:

---

[145] A detailed presentation of the structure and operation of the calibration system of the MAGIC cameras can be found in GAUG (2006).
[146] Credit of the picture: Robert Wagner.

(5) The «Counting House»
To minimize the weight of the structure, the electronic components of the camera (the trigger and readout electronics) are not located in the camera itself, but in a nearby building: the «counting house». The transmission of signals from the camera to the «counting house» is done via optical fibres, because this way the transmission remains largely free of background noise. For this purpose, the electrical impulses from the PMTs are first converted into optical pulses, and in the «counting house» this optical signal is replaced by an electrical signal.

The processing of the data in the «counting house» is different for MAGIC I and II, and over the years, even the processing technique for MAGIC I data was modified once. But on the whole one can summarize what happens in the «counting house» as follows: the incoming signal is split. One part passes to the trigger system and the other part to the «data acquisition system». There it is digitized and temporarily written in a ring buffer. If the trigger system classifies the information as relevant, the appropriate part of the ring buffer is written on a disk.

Finally, the raw data gathered that way is sent to the MAGIC Computing Center in Barcelona («port d'Informació Científica») and Würzburg («Würzburg DATA Center») via post and/or internet, where the data are made available for the data analysis, while another further copy of the data remains stored as backup copy in La Palma.

Let us now try to gain some information for our discussion from the preceding report. Does the outlined structure of the MAGIC telescopes reveal something about the possibility or impossibility of a theory-driven control of measuring instruments in astroparticle physics?

Can we state, for example, based on Collins and Pinch and this report, that there are no formal criteria to determine whether an experimental device, such as MAGIC, is working properly or not, and that the answer to this question represents only the result of a negotiation between specialists? Or can we reject the position of Collins and other constructivists on the basis of MAGIC?

I think it is possible to propose an argument against the constructivist view of the measuring devices as plastic resources on the basis of the previous description of the MAGIC telescopes. But the argument, which can be formulated at this point, is merely the same argument which has already been discussed in the previous chapter and also in the introduction to this chapter. That it is the same argument is not surprising, as the MAGIC telescopes have so far been described on a detail level that is comparable with the descriptions in the last chapter. Therefore, the discussion of the theoretical control of the measuring devices cannot be continued at this point by adding new details about the severity of the development, deployment and calibration procedures of the MAGIC Telescopes.

So how can the argument against the constructivist standpoint concerning the measuring instruments become more plausible?

What can be said at this point is that the MAGIC telescopes consist of individual parts, whose theoretical foundations and function mode are well known. And that means, that we have learned to verify that these components work correctly due to many scientific and commercial applications:

The basis plates for the mirrors, for example, were produced by the company *Compositex*[147] (in Vincenza, Italy), a company specialized in the production of all kinds of structures made of very light carbon fibres for engines, aircraft, ships, etc. The protective coating of the mirror was produced by *Fraunhofer-Institut für Fertigungstechnik und Angewandte Materialforschung*[148] (Bremen, a research institute whose researchers have a great deal of experience and broad expertise in the field «adhesion and interfaces». The diodes for the calibration system were purchased from the Japanese company *Nichia*[149], specialized in the manufacture of light-emitting and laser diodes and other lighting systems for commercial purposes, etc. The properties of the components were first tested by their manufacturers, and further tests were then carried out by the responsible groups within the MAGIC Collaboration. It is, indeed, not easy to see how much

---

[147] http://www.compositex.it/eng/azienda/index.php
[148] http://www.ifam.fraunhofer.de/
[149] http://www.nichia.com/de/about_nichia/index.html

free space will be left for theory-based interpretations – of the functioning of the telescope components.

But while these remarks emphasize an important point –namely, that the proper functioning of the individual parts of the MAGIC telescopes (and more generally the elements of the detectors in astroparticle physics) are not arbitrarily interpretable–, such a finding is not sufficient to exclude the possibility that theoretically desired types of raw data can be generated in the telescopes by a skilful composition of their components.

Because Collins and Pinch or other like-minded constructivists could counter at this point that, for example; both, the devices with which Weber claimed to have detected gravitational waves, and the devices with which Weber's opponents argued that such amounts of gravitational waves do not exist and that they consist of individual parts whose proper functioning is clear[150]. Could it not also be the case, that the performance of the basic elements of the MAGIC Telescopes is clearly defined, but at the end the whole construction supplies the data needed for the confirmation of certain theoretical preconceptions?

This constructivist objection can be further strengthened, when one considers that, although the basic components of the telescopes have been acquired «ready-made», however, a significant proportion of these components has been used for the development of new devices and systems which are to be regarded as genuine technical developments. Devices such as the cameras or systems like, for example, the *calibration system*, or the *data acquisition system* of MAGIC are new, even if they consist of known components. How can one be certain that they only play an imaging rather than a constructing role?

Furthermore, it does not seem satisfactory to constantly mention that the parts of the various instruments are known and tested, without giving any examples of the different types of these tests.

In the next subsection I will now try to remedy the shortcomings of the above mentioned argument against the constructivist view of the measuring devices as plastic resources. For this purpose some tests must be outlined to ensure the

---

[150] The controversy about the alleged discovery of gravitational waves by Joseph Weber appears to the constructivists as a prime example that it is impossible to determine objectively whether a new device is working properly or not. According to Collins and Pinch this question can only be decided by the «power of persuasion» of some scientists. And this implies at the end that the specialists consider as «well-functioning equipment» those instruments that generate the data that are adapted to certain theoretical preconceptions. Collins and Pinch have for example in their discussion of the «affair Weber» suggested that an important factor that moved the specialists to regard the devices of Weber's critics as correctly functioning, was the fact that they provided results concordant with the current cosmological scenario (in contrast to Weber's results). To the controversy about the experiments of Weber see COLLINS - PINCH (1998) chapter 5, and FRANKLIN (1999) chapter 8.

correct functioning of the parts of the MAGIC Telescopes. In addition, some of the steps taken during the commissioning of MAGIC, to ensure the correct functioning of the telescopes as a whole must be briefly described.

3.2 Tests of MAGIC

The production of the components of the first MAGIC telescope began in 1999. It is difficult to determine the exact number of different tests that have been carried out since then with the purpose to check the performance of the various parts and systems of the telescope and the telescope as a whole. This number of tests certainly exceeds one hundred; and we are just talking about the number of test types and not about the number of routine checks, which naturally is several scales higher.

Needless to say, that in this subsection only a small sample from this great variety of control measures can be presented. However, we shall pay particular attention to the fact that in this sample three different groups of tests are represented. These are the following: (1) checks of the performance of individual parts of the telescope, (2) checks in connection with the updates of the systems, and (3) checks to monitor the correct performance of the telescope.

From the first group, we will deal here more thoroughly with some tests to supervise the performance of the mirror and to check the laser for the optical transmission of signals from the camera to the «counting house». We will briefly discuss some tests from the second group in connection with the update of the *data acquisition system* and the camera. And we will consider some steps from the third group that were taken during the commissioning of the MAGIC telescope. For reasons of simplicity, most examples refer to MAGIC I. Exceptions are the tests for checking the mirrors, which were extracted from an investigation in connection with the development of MAGIC II. Further exceptions certainly are the tests in connection with the development of a new camera, which can be relevant for a future update of both telescopes.

3.2.1 Tests to Check the Mirrors of MAGIC II

After the production of the mirrors of MAGIC II, several quality controls were performed to verify whether they meet a number of requirements. The key requirements for the mirrors are the following:

1. The mirror elements must be weatherproof.
2. The *point spread function* [PSF] of the mirrors must be smaller than the diameter of a camera pixel (that is ~ 24 mm long).

3. The reflectivity of the mirrors should be over 85% within the wavelength of 320 nm to 600 nm, and at least 70% in the range from 280 nm to 320 nm. (These are the relevant wavelengths to optimize the detection of Cherenkov light).

Here only the tests will be described, which refer to the *point spreading function*. The reader interested in further details referring to the tests of the MAGIC Mirrors should read the third chapter of the thesis of Cornelia Schulz –«Novel All-Aluminium Mirrors of the MAGIC Telescope Project and Low Light Level Silicon Photo-Multiplier Sensors for Future Telescopes»–.

For checking the *point spread function* [PSF] of the mirror a «*charge coupled device*» (CCD) camera ST-402ME from *Santa Barbara Instrument Group* was used. This camera was the main component of the following experimental assembly[151]:

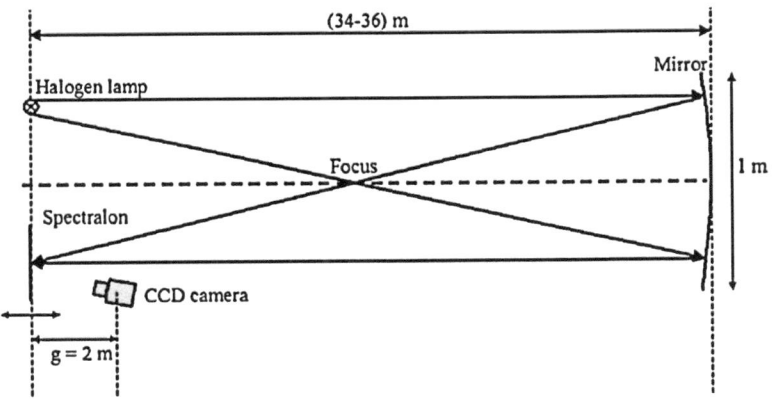

The arrangement and the test procedure for determining the *point spread function* of the mirrors were described as follows by Cornelia Schulz:

«A CCD Camera is set at about 2 m distance from the target screen perpendicular to the imaging spot on the front surface of the Spectralon. This distance g is also measured. The reflected image is then recorded with the CCD camera and also a background image for which the halogen lamp is screened. [...]

At first, the center of the light spot image is determined by assuming a Gaussian spot shape for the horizontal and the vertical intensity distribution. Within a sufficiently wide integration window the intensity is computed by adding up the intensity recorded by each pixel. Then, a radial integration in

---

[151] Courtesy of Cornelia Schultz

small steps in r is done, beginning from the determined spot center. The radial distance to the spot center is increased incrementally by one pixel and the integral of the light intensity is evaluated. [...]

If the integral equals nearly 90 % of the total signal the iterative integration is repeated once more. The respective R90 is calculated by interpolating between the previous and the actual radius. In this way the error of R90 can be slightly reduced. Since the unit of the evaluated R90 is defined by the number of pixels, it is finally converted to a metric distance. Therefore, the measured distance of the CCD camera to the Spectralon screen is used for computing the actual size of one pixel. This is done by using the imaging equation for a thin lens»[152]

Although this method is very reliable, an independent check of the results was still carried out. This check was realized by the application of a second method for the calculation of the PSF:

«In order to verify the correctness of the calculated pixel size, this value is compared to the pixel size that is determined by another procedure. A paper with a known grid of paper is placed on the Spectralon and recorded by the camera. From the size of one square of the grid, which is 5 x 5mm, one can calculate the pixel size. This is done by counting how many pixels fit along the lateral length of one square. The ratio of the number of pixels Δx and the lateral length (5 mm) equals the metric pixel size. The above mentioned further calculation is more precise than this evaluation; therefore this method is only used for a crosscheck.»[153]

This is an example of a practice used for all technical checks in connection with the diverse components of the telescopes: the «cross check». That is, one always tries to use different and independent methods for the computation of the parameter whose value we want to determine[154]. In an ideal situation the different methods should be comparable with regard to accuracy. But if this is not possible the alternative methods must be at least so accurate that the consistency between

---

[152] SCHULZ (2008).

[153] Ibidem

[154] As another example of cross checks in the context PSF of the MAGIC mirrors, can be cited the tests, which were carried out in the first months after the inauguration of MAGIC I, to determine the PSF of the entire reflector:
«Muons generate very well defined rings in the camera that are used to calibrate the telescope and to derive the point spread function of the reflector. [...] The point spread function can also be derived by means of the distribution of pixel anode currents of bright stars". CORTINA (2004).

the different results can be used as proof of the correctness of the results obtained with the theoretically best method.

3.2.2 Checks to Test the Operability of the Lasers for Optical Transmission of Signals from the Camera to the «Counting House».

In the previous subsection it was mentioned that the transfer of the signal from the camera to the «counting house» takes place through optical fibres, mainly because this way, the transmission remains free of background noise.

For this kind of the transmission it is, however, necessary to convert first into optical impulses the electrical impulses which are generated in the photomultipliers of the camera. After investigating various alternatives, it was decided that this step should be realized by lasers of the type VCSEL [«vertical cavity surface emitting laser»].

However, these lasers were relatively new at the time of the construction of the MAGIC I Telescope. They came onto the market in 1997 and were so far only used for the transmission of digital signals, while the signal of the MAGIC Camera should be transmitted in the analogue mode. Therefore, it was necessary to clarify various aspects of the functioning of the VCSEL, before the laser could be used.

To this end, the following experimental assembly was set up:[155]

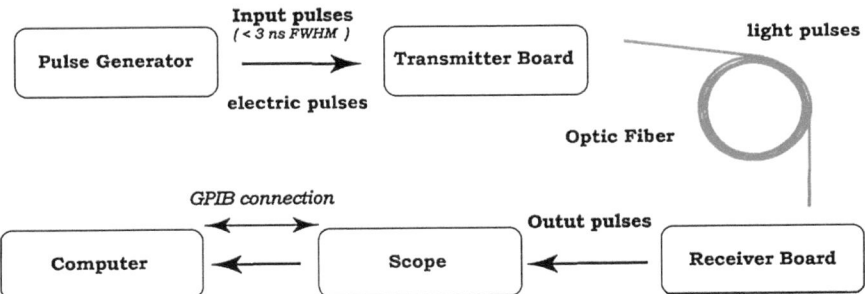

Paneque Camarero who was actively involved in the development of this aspect of the MAGIC technology, described the arrangement as follows:

> «We use a special pulse generator to inject PMT like pulses (≤3 ns FWHM) of preset amplitudes into the transmitter board, where the VCSEL transforms these signals into light pulses that are sent via optical fibres to the receiver board. In the receiver board, a GaAs PIN-diode converts the light pulses back

---

[155] See PANEQUE CAMARERO (2004) 132. Courtesy of the author.

into electric pulses, which then are read by a digital oscilloscope that is controlled by a PC through a GPIB connection.»[156]

With the help of this experimental assembly it is possible to examine the behavior of the VCSEL:

«We used this setup to characterize our optical link system by studying both the amplitude and the area of the output pulses vs. time and also vs. the amplitude of the input pulses. [...]

The electronic components (transistors, operational amplifiers...) and the PIN-diodes used in the optical link system are well known to be of low noise and stable. Therefore, the measured noise and stability of the output pulses reflects, in first order, the performance of the VCSELs.»[157]

Some important information about the function mode of the lasers was gained this way. The most important are the following:

«[A] study of the measured amplitude and area of the output pulse vs. time showed that some VCSELs can be quite noisy. [...]

We found that the relative noise in the output pulse increases when decreasing the amplitude of the input pulse, and that the forward current applied to the diode (bias current) can affect this noise too.»[158]

In order to guarantee that this would not lead to a distortion of the measurement results of MAGIC, some steps were taken:

«• The signal amplification in the transmitter board was increased by a factor four in order to modulate the light output of the VCSELs by a four times larger pulse. This modification reduces the relative noise of the transferred signals. The performance improvement is particularly important when transferring low amplitude pulses, which (due to the exponential decrease of the fluxes of γ - rays and hadrons with the energy) will be the case for most of the observed atmospheric showers.

• The use of 6 mA bias current to drive the VCSELs. This modification reduces the number and the relative intensity of the "resonances" and the sudden changes in VCSEL gain. This current is high enough to ensure that all diodes are working in the lasing mode, i.e, well above their threshold currents [...]. We decided to use only one bias current for all VCSELs in order to sim-

---

[156] Ibidem 129.
[157] Ibidem 131.
[158] Ibidem 133; 135.

plify the replacement of dead or malfunctioning lasers in the telescope camera.

In addition, due to the significant performance spread among the measured VCSELs, detailed quality checks were carried out for each single VCSEL; and all those lasers not fulfilling the strict requirements to be used in MAGIC were rejected.»[159]

The quality checks performed for each VCSEL can be summarized as follows:

«The selection of the VCSELs was carried out in two steps. In the first one, the performance of the laser diodes was studied during short periods of time at several bias currents settings, ranging from 5 mA to 7 mA in steps of 0.05 mA. This is the so-called bias current scan test. The VCSELs were accepted/rejected according to the measured performance close to 6.0 mA, which is the selected working point [...]. Those VCSELs that passed the first test were submitted to the second test, in which their performance was studied at a bias current of 6.0 mA during a time period longer than 10 hours; this is the so-called long time test. Both tests were performed using the standard input pulse defined in the previous section.»[160]

As result of these tests, 670 1out of 970 were selected for their use in the MAGIC Telescope.

3.3.3 Tests related to the update of the «Data Acquisition System» of MAGIC I

In the previous subsection it was pointed out that the analogue signals transferred from the camera of the telescope to the «counting house» are split there first and then digitized by the «data acquisition system». Digitization is carried out by the so-called FADCs [«flash analog-to-digital converters»]. When MAGIC I was developed, it was used for digitizing the cheap but relatively slow *Siegen–FADCs*. The problem with this decision was that the *Siegen–FADCs* only can generate 300M samples/s. For a digitization of the gamma events, which cause Cherenkov- flashes lasting only a few nanoseconds, this digitization rate is not optimal. Because the pictures it supplies of the gamma events are not detailed enough and it, therefore, makes it more difficult to separate both, the signal and background light and the gamma and hadron events.

These problems could be solved by installing faster FADCs. But the fast FADCs are so expensive that this solution was not possible. For this reason, some members of the MAGIC Collaboration developed a new digital system

---

[159] Ibidem 138.
[160] Ibidem 140.

based on the «Fibre-Optic Multiplexing» technology (MUX). This update of the «data acquisition system», allowed a better time resolution of the events –since the new digitization rate is 2G samples/s– and, thus, the sensitivity of MAGIC was clearly improved. Due to lack of space the details of the new system are ignored here[161]. However, it is instructive for our discussion to mention, which steps have been taken to ensure that the MUX FADCs work correctly:

First, some pixels of the MAGIC camera were digitized using the new system. In addition, the expected signal of these pixels could be determined by interpolation of the signal of the neighbour pixels. And this way it was tested whether the results of the new digitization were consistent with the results of the old and well established digitizing procedure[162].

Afterwards the system was completely installed without the old system being switched off. Instead the analogue signal in the «counting house» was split between April 2006 and January 2007, and so for months both systems were running parallel for further testing.

During this period the old and the new digitization of the signal from the Crab Nebula were especially compared. Because the emission pattern of the Crab Nebula is well understood, this source is usually used as a calibration source[163]. In January 2007, the old FADC system was finally switched off.

3.3.4 Tests in Relation to the Update of the Camera

In the previous chapter it was pointed out that one of the main difficulties with which the Cherenkov telescopes are confronted, is not only the short duration but also the very weak intensity of the Cherenkov flashes which are to be detected. To overcome this unfavourable factor it is, among other things, necessary to install extremely sensitive photosensors in the camera of the telescope.

When the MAGIC Telescopes were constructed the well-known PMTs, mentioned in the previous subsection –in the description of the camera–, were used for this purpose. But these photomultipliers are heavy and fragile, and they reach a photon-detection-efficiency (PDE) of only 20 to 30%. That is why the MAGIC collaboration are now focussing on the development of a new camera based on innovative photodetectors.

For the construction of new detectors, the technology of the «*Geiger-mode avalanche photodioden*» (GAPD) is now tested; a technology currently being developed. For the lack of space the details of the new detectors must also be ignored[164]. It is, however, interesting to list the advantages and disadvantages of

---

[161] The interested reader is recommended the following paper: GOEBEL et al. (2007).
[162] See BARTKO (2005).
[163] See TESCARO et al. (2007).
[164] Some details are described e.g. in OTTE et al. (2007).

the GAPD compared to the PMTs. Otte et al. summarized these aspects as follows:

> «The main advantage of G-APDs employed in astroparticle physics is the potential high PDE which can be up to three times as high as that of classical PMTs. Other advantages are:
>
> – single photoelectron response
>
> – compactness
>
> – insensitivity to magnetic fields
>
> – low operation voltage (< 100 V) and high gains (10^5 - 10^6)
>
> – no damage when exposed to sunlight, even when under bias
>
> – long-term prospects for low fabrication costs.
>
> Disadvantages of G-APDs are:
>
> – current sensors sizes are limited to < 10 mm^2
>
> – dark noise rates between 100 kHz and several MHz per mm^2 sensor area at room temperature
>
> – so-called optical crosstalk, i.e. correlated firing of several cells»[165]

It is important to refer to the possible disadvantages of the new technology (as done in the above quotation). Because this clearly shows that GAPD cannot be used without prior thorough testing.

What tests have been performed by the members of the MAGIC collaboration so far?

First, a GAPD pixel was built into the current MAGIC camera. And its performance was compared with the performance of the adjacent pixels by observing various particle showers[166]. The test led to the following results:

> «– It is possible to detect Cherenkov light from air showers with available G-APDs.
>
> – With the tested G-APDs, we obtained a significant increase in the efficiency to collect Cherenkov photons from air showers compared to a MAGIC PMT-pixel ($\sim 60\%$).
>
> – The intrinsic noise rate of the tested G-APDs at room temperature was a factor of 10 below the level of the night sky background light.»[167]

---

[165] Ibidem.
[166] Ibidem.
[167] Ibidem.

The fact that the photon detection efficiency (PDE) of the GAPD pixel was 60% higher than the standard PMT pixels –after the calculation of all possible error factors, in particular the correlated activation of neighbouring cells–, was interpreted as a promising result which showed that the new technique could potentially be able to provide for a abrupt improvement in the efficiency of the telescopes.

Nevertheless, it was decided to do further tests. Then a 36-pixel camera was built and tested (as prototype) that uses only GAPD pixels[168]. The camera was placed in the focus of a spherical mirror with a diameter of 90 cm. In the night of the 9th July 2009 the night sky was observed with this device in zenith direction for about an hour. This test was carried out in Zurich, and this implies that the observations took place under very adverse conditions due to the strong background light. Nevertheless, it was shown that even under these conditions the new camera was able to detect Cherenkov light in particle showers.

The development process of the new MAGIC camera, thus, is not yet complete. Now the plan is to build a small imaging air Cherenkov telescope in the vicinity of the MAGIC telescopes, which is to be endowed with a GAPD camera. If the tests with this small telescope –provisionally called «FACT»– are successful, the camera of the MAGIC telescopes might be replaced by a GAPD camera.

3.3.5 Testing the Proper Function of the Telescope

So far we have only presented some tests to check the function of individual components of the MAGIC telescopes. However, it is of particular importance to examine whether the composition of these components leads to a properly functioning system –the whole telescope –. How can this examination of the whole telescope be done?

The usual procedure for many of the tools of experimental physics is to use the new instrument to measure signals with known physical properties. The results of these measurements [outputs] are then compared with the well-known values of the physical parameters of the measured signals [input]. And this way, the new instrument is calibrated[169].

But this procedure is unfortunately not applicable with regard to the instruments of astroparticle physics. Because one can e.g. send light pulses to check the function mode of a camera –and that is exactly what the calibration system of the MAGIC Telescopes does –, but one cannot produce an artificial source of high-energy gamma photons for an absolute calibration of the

---

[168] See ANDERHUB (2009).

[169] One finds an excellent analysis of the philosophical safety of this procedure from a realistic point of view-in FRANKLIN (1999) chap.8.

telescopes. Luckily there is, however, a natural source, whose characteristics were very precisely investigated in different experiments (Cherenkov Telescopes and Satellites). It is the Crab Nebula, a pulsar that has a very constant and strong emission of gamma rays and which was detected more than 20 years ago[170], and which has been constantly observed since then. The spectrum of this source is thus well known, and therefore, the Crab Nebula is usually used as a calibration source for new detectors of gamma radiation:

> «The Crab Nebula is a steady emitter at GeV and TeV energies. The γ-ray emission is produced by Inverse Compton (IC) scattering of a population of electrons that are accelerated in the plerion around the central pulsar. The spectrum of this source has been measured in the GeV range by EGRET and at energies above 300 GeV by a number of Cherenkov telescopes. The expected level of emission between 30 and 300 GeV makes it into an excellent calibration candle for MAGIC.»[171]

For that reason the Crab Nebula was observed as the first source at the beginning of 2004, when the first MAGIC Telescope was still in testing:

> «The γ rate from the Crab Nebula can be used to give a very rough estimate of the threshold energy of the telescope. The idea is to compare the measured γ rate with that expected from the known γ-ray flux of the Crab and the efficiency of the MAGIC Telescope to detect γ-rays. This estimation is based on two assumptions: a) the flux of the Crab at low energies (~ 100 GeV) can be extrapolated from the measured flux of the Crab Nebula at energies above 300 GeV, and b) the Monte-Carlo-simulations describe in first order the experimental data».[172]

On 14. and 15. February 2004 Paneque Camarero made the first observations of both, the Crab Nebula and the active galaxy Markarian 421. The reason for the selection of these sources was, that the Whipple Telescope and other detectors (of both, gamma and X-ray radiation) had reported a flare of the active galaxy at that time. Therefore, one used this opportunity to test whether this flare could be confirmed by the new telescope. The results of these observations were very positive:

> «The detected γ rate measured for the Crab Nebula is about half the one we measured for Mkn 421. However, the γ-ray flux of Mkn 421 is usually an order of magnitude lower than that of the Crab Nebula. Therefore, we concluded that Mkn 421 was in a strong flaring state during the night of 14th-

---

[170] WEEKES et al. (1989) 379-395.
[171] CORTINA (2004).
[172] See PANEQUE CAMARERO (2004) 230.

15th February. It is worth mentioning that the increase in the photon flux from Mkn 421, during February 2004, had been already observed in the very high energy γ - ray domain (≥ 300 GeV) by the WHIPPLE Telescope, and also in the X-ray domain (2-10 keV) by the All Sky Monitor detector (ASM) borne on the Rossi X-ray Timing Explorer satellite (RXTE).»[173]

In the following months similar tests were done that eventually convinced the experts that the first MAGIC telescope is working properly[174]. This conviction is based primarily on the fact that the telescope produces results that are consistent with results of other (already established) detectors. There also are other kinds of tests which delivered independent evidence for the correct operation of the telescope. The analysis e.g. of the muon background in data recording provided additional evidence of the correct functioning of MAGIC. Because the atmospheric muons cause a very characteristic trace in the camera of the imaging Cherenkov telescopes. By means of computations and simulations a theoretical study was made about the characteristics of this muon background, and the analysis of the recorded data showed congruency with such theoretical expectations:

«Measuring muon rings and arcs allows for the calibration of an Imaging Air Cherenkov Telescope, since the geometry of the muon track can be reconstructed by the image in the camera. We extracted 0.5% muon events from the data obtained during commissioning of the MAGIC telescope in the summer of 2004, corresponding to a rate of 1Hz. The distributions of the observed muon parameters agree well with the simulated ones.»[175]

It does not seem to be necessary to mention here additional examples of tests for checking both the development and commissioning of the MAGIC Telescope. Instead, we should try to extract some notes for our discussion from the previous examples. The question we must now examine is: what factors determine the belief of experts that a new device is working properly (or not)?

There is a constructivist and a realistic answer to this question. According to the constructivist view, factors such as the advantages and disadvantages for the participating research groups, experts' confidence in the expertise of designers, various methods of propaganda, the prestige of the participating institutions and

---

[173] Ibidem 228.

[174] To this point writes e.g. Cortina: «In the months elapsed since the inauguration of the telescope, we have mainly concentrated on low zenith angle (<40) observations of standard TeV candles like Crab Nebula, Mrk 421, Mrk 501, 1ES 1426 and 1ES 1959. [...] We have selected only very short data samples for which the weather conditions were excellent on the basis of the trigger rates and star extinction measurements» CORTINA, J. (2004).

[175] MEYER - MASE (2004).

collaborations that promote the new development, etc., are the determinants for decisions about when a new device is considered fully operational[176]. And in this context it is obvious that a theoretical proposal that has come into fashion, can indirectly contribute to confirm itself already at the level of designing research instruments. Because the confidence of scientists in the new theoretical scenario will lead to different ways to ensure that the instruments are considered only to be properly working, when they provide the desired data.

In contrast to this, the realistic view attributes the confidence of experts that a new device is working properly results, above all, to the different types of tests, which were made for the examination of the function mode of both, the individual components and the equipment as a whole. These tests leave little (or no) space for theories directing the operation of the instruments.

For the acceptance of a new device from the perspective of constructivist, thus, mainly sociological reasons are important, while the realists consider the technical aspects as the truly decisive reasons.

Certainly we cannot settle such a controversy for each individual case in the context of this study. But with regard to the MAGIC Telescopes, we can state that the confidence of the experts, that these telescopes provide correct data is supported by an impressive number of technical controls. That is, the realistic perspective seems to supply –at least at this point– a satisfactory understanding of the practice of physicists.

However, a constructivist could argue that a detector such as MAGIC is not the appropriate instrument, in order to show the social control of the correctly functioning devices. Because this control based on factors such as prestige, rhetoric or benefit, takes place only (or mainly) in experiments where a new phenomenon is investigated with a new device. Only in this case occurs

---

[176] Collins and Pinch have e.g. listed a whole inventory of non-technical reasons that were determinants for the decision whether the various experiments in the discussion of Weber's experiments gave correct results (or not):
«1. Faith in a scientist's experimental capabilities and honesty, based on a previous working partnership.
2. The personality and intelligence of experimenters.
3. A scientist's reputation gained in running a huge lab.
4. Whether or not the scientist worked in industry or academia.
5. A scientist's previous history of failures.
6. "Insider information".
7. Scientist's style and presentation of results.
8. Scientist's "psychological approach" to experiment.
9. The size and prestige of the scientist's university of origin.
10. The scientist's degree of integration into various scientific networks.
11. The scientist's nationality» COLLINS - PINCH (1998), *The Golem. What you should know about science* (Cambridge, Cambridge University Press 1998) 101.

something that Collins and Pinch have described as the «experimenters' regress». An «experimenters' regress», which can only be broken through a near-consensus between the specialists, and the acquisition of which would mean that the representational function of nature, the realists associate to the natural sciences, is lost.

What is the «experimenters' regress» which is supposed to make the realistic understanding of the natural sciences impossible? Collins and Pinch explain this with the help of their favourite example, the experiments of Weber regarding the existence of gravitational waves:

> «The student can have a good idea whether or not he or she has done an experiment competently by referring to the outcome. If the outcome is in the right range, then the experiment has been done about right, but if the outcome is in the wrong range, the something has gone wrong. In the real time, the question for difficult science, [...] is, *"What is the correct outcome?"* Clearly, knowledge of the correct outcome cannot provide the answer. Is the correct outcome the detection of gravity waves or the non-detection of gravity waves? [...]
>
> Thus, what the correct outcome is depends upon whether there are, or are not, gravity waves hitting the earth in detectable fluxes. To find this out we must build a good gravity wave detector and have a look. But we won't know if we have built a good detector until we have tried it and obtained the correct outcome. But we don't know what the correct outcome is until... and so on *ad infinitum*.
>
> This circle can be called the "experimenter's regress". Experimental work can only be used as a test of some way is found of breaking into the circle of the experimenter's regress. In most science the circle is broken because the appropriate range of outcomes is known at the outset. This provides a universally agreed criterion of experimental quality. Where such a clear criterion is not available, the experimenter's regress can only be avoided by finding some other means of defining the quality of an experiment; and the criterion must be independent of the output of the experiment itself.»[177]

And exactly in these pioneer works at the limits of research, an invention accepted by consensus replaces (after the constructivist view) the description of nature. These are certainly very special circumstances. But the «constructivist» decisions made under these circumstances give the whole research process the

---

[177] Ibidem 97-98.

character of a pure invention, in no way determined by the mode of being of nature[178].

For our case this has the following consequences: the tests to verify the correct function mode of the MAGIC telescopes, mentioned above, have surely been designed in the best experimentalist tradition. But they depend on the condition that the Crab Nebula serves as a standard candle for the calibration of Cherenkov telescopes. However, the first imaging air Cherenkov telescope could not be calibrated by this source, since this telescope measured the gamma flux of the Crab Nebula and discovered the non-variability of this flux for the first time. How could the correct function of this first Cherenkov telescope have been tested?

Now we try to answer this question. But first, I would like to stress that not only the results of the previous Cherenkov telescopes were used as support for the calibration of MAGIC, but also the above-mentioned satellite data of EGRET. That is, the values of the gamma flux from the Crab Nebula which helped to calibrate the measurements of MAGIC, originate from two different (and from each other independent) kinds of experiment. A fact that undoubtedly increases the reliability of these values.

---

[178] I have already noted (in paragraph 5 of the first chapter) that Collins seems to emphasize in his later works the role of rational arguments in this decision-making. But there remain in these texts a number of references to economic and academic interests that influence the decisions. The result of this mix of rational arguments and particular interests is ambiguous: Do such interests contribute simply to reinforce decisions that also would have been taken anyway by rational arguments? Or do such interests remain the key determinants for something being interpreted or not interpreted as a discovery?
In the latter case, nature continues having little to say in the conception of Collins. And indeed, the metaphor that Collins recently presented to illustrate their position seems to point to this. We found this metaphor in "Gravity's Ghost":
«It is not Nature, reflected clean and pure on the still surface of science's pool; there is no pool, only the fast flowing river if human activity on a restless Earth. The first detection will be a social and historical eddy...» Collins (2011) p.20-21.
This text is certainly poetic. And as usual with the poetic texts, it may receive different readings. But I think that words as "pool" and "reflected" are probably referring to the usual realistic account of the scientific theories as descriptions of the mode of being of nature. And this view is being denied. While expressions such as "social and historical eddy" quite possibly refer to a constructivist view of phenomena as cultural products resulting from a confluence of argumentations and interests, in which nature (as something "out there") hardly plays a very secondary role.
If this interpretation is correct, I think that the position of Collins has not changed substantially. (And this is not the most radical interpretation at all. Alternatively, one can interpret the metaphor of Collins as the assessment that there is no nature "out there" but only a "fast flowing river").

But let us forget this detail now, and focus on the question of how it could be established, if the first Cherenkov telescope (with which one observed gamma-ray sources like the Crab Nebula for the first time) worked correctly. Fortunately, we do not have to speculate a lot about this issue, because all the relevant tests are summarized in the paper of Weekes et al. «Observation of the TeV Gamma Rays from the Crab Nebula Using the Atmospheric Cherenkov Imaging Technique»[179].

The instrument, with which gamma radiation from the Crab Nebula was first clearly observed is the WHIPPLE Telescope. The first observations were carried out in the years 1986-1988, and the results were published in 1989 in the aforementioned article. This paper, therefore, is regarded as the birth of the current Cherenkov technique for study of the cosmic sources of gamma radiation.

It is not necessary to go into the details of the WHIPPLE telescope. For the design of this telescope basically is the same as that of the MAGIC telescope[180]. And that means; that the tests to check the functionality of WHIPPLE could easily have taken place at the start of MAGIC... when no calibration data was available.

How did the members of the WHIPPLE Collaboration become so confident that their instrument supplies correct results? The answer to this question is: through consistency checks of the results. That is, they regarded the obtained results from different perspectives, in order to examine whether all details were consistent with each other. In other words, they have derived the plausibility of the results of WHIPPLE from the features of the data itself. The consistency checks were carried out as follows:

«Since this is the first demonstration of the power of the imaging technique, we look for further evidence that the results are internally consistent with the detection of a flux of gamma rays. We have already shown that (i) apart from the differences in the gamma-ray domain, the ON and OFF distributions are similar; (ii) the net excess that is seen in *azwidth* is seen also in the two parameters that are the basis of this selection, *width* and *miss*, thus demonstrating that both shape and orientation are effective in selecting gamma rays: (iii) the excess is seen with the other parameter cuts also; in particular, we see the same strong effect when we use the four/six cut combination outlined by Hillas (1985); (iv) the excess is constant with time and evenly distributed over

---

[179] WEEKES et al. (1989).
[180] Of course, many elements of the telescope WHIPPLE are less developed than the corresponding elements of MAGIC, because it is an instrument that was built more than a decade before the MAGIC telescopes. (The camera used by WHIPPLE at this time had for example only 37 pixels). But these details are not relevant for the following discussion.

the 175 ON/OFF pairs; (v) there is evidence in an independent earlier data set for the same emission.

As further tests for possible systematic effects, we check the following: (vi) if the excess were noise-generated, it should be most apparent in the events that are just above the threshold; division of the data by total digital counts recorded should show that the effect is not dominated by the least bright images; (vii) discrimination is less efficient at larger zenith angles; division of the data base by zenith angle should show that the technique is mist sensitive close to the zenith; (viii) the effect of the star, Zeta Tau, should be shown to have no influence on the event selection; (ix) the order of observations (ON before OFF) should be shown to have no influence on the result»[181]

It is not necessary to deal with all the terms used in this quote. In order to form an idea of the argumentation strategy conceived by the members of the WHIPPLE Collaboration, it is sufficient to explain that *azwidth*, *width* and *miss* describe features of the different camera pictures; that in ON-observation mode the telescope focuses on the direction of the investigated source while in the OFF-observation mode the telescope moves in any direction, and that the star *Zeta Tau* is a neighbouring star of the Crab Nebula .

However, it is important to understand the argument that is behind all the technical details listed in the quote. This argument can be summarized as follows: if the detected signal is a genuine signal and not an artefact produced by the telescope, it is expected that the data will have a certain characteristics. If one determines that all these characteristics apply –as in the case of the observations of the Crab Nebula by Weekes et al.–, one can consider the assumption that the detected signal is not an artefact as convincing. Of course, this is not a logical proof, and it is unnecessary to point out that the experiments of physics sometimes lead to wrong conclusions. Because the properties of the data could possibly be explained differently. But the more details of the data are discovered, which can be explained by assuming that the signal sought is really there, the more credible it appears that you have detected a genuine signal.

This way of thinking corresponds with Franklin's fourth strategy to confirm the realism of the results of experiments, presented by Franklin in its essay «Experiment in Physics»:

> «Using the results themselves to argue for their validity. Consider the problem of Galileo's telescopic observations of the moons of Jupiter. Although one might very well believe that his primitive, early telescope might have produced spurious spots of light, it is extremely implausible that the telescope would create images that they would appear to be a eclipses and

---

[181] WEEKES et al. (1989) 390.

other phenomena consistent with the motions of a small planetary system. It would have been even more implausible to believe that the created spots would satisfy Kepler's Third Law (R^3/T^2 = constant).»[182]

There are strong parallels between Franklin's example (the telescope of Galileo) and WHIPPLE. Galileo's instrument was, as Whipple, a new kind of instrument[183]. Both instruments have explored new phenomena. Therefore, the question in both cases could be, if the new phenomena were real, or just artifacts produced by the apparatus. But in both cases, the properties of the discovered phenomena were in contrast to the possibility of such delusion. The regularity of the movements of the satellites of Jupiter, and the fact that they were describable as planetary movements, strongly suggested the real existence of new celestial bodies orbiting around Jupiter, as Galileo correctly assumed. And the temporal stability of the results of WHIPPLE and the stability of these results where the event segment with the most dominating background effects is concerned, as well as the stability of the results for changes in the parameters for the selection of the gamma events or for changes in the trigger logic etc., strongly suggested that a source of gamma radiation has actually been detected.

It was, therefore, justified to recognize the Whipple telescope as a well-working equipment –without having to fall back on rhetoric, propaganda, social policy or other mechanisms suggested by the sociologists to break the «experimenters' regress»–, only based on rational grounds. And so it can be said that the tests used in order to check the operation of the MAGIC telescopes, for which the results of WHIPPLE and other Cherenkov telescopes were used, are legitimate and satisfactory from a realistic point of view.

In summary it can be stated that regarding the question, whether a measuring instrument such as MAGIC or WHIPPLE or other instruments of astroparticle physics work properly or not, there is little up to no space for interpretation. The tests to monitor the function and performance of these instruments are convincing, and there is no infinite «experimenters' regress», which would have to be broken with constructivist interventions.

The measuring instruments as such, therefore, cannot be the key figures for a control of the experiments by theories. Does this control, thus, happen on the level of the data analysis? We will examine this possibility in the next section.

---

[182] FRANKLIN (2009).
[183] There were different away-preparing attempts, on which the Design of WHIPPLE based. Since however none of the previous experiments could detect a gamma source clearly, we can call WHIPPLE with some justification the «first» imaging air Cherenkov telescope.

4. The Analysis Chain of MAGIC

The data recorded by means of the MAGIC Telescopes and stored in Barcelona, Würzburg and La Palma only contain indirect information about the observed sources of γ-radiation due to several reasons. This is the case just because the Cherenkov light, one hopes to detect with t these devices, is not directly caused by the primary gamma photons but by the secondary particles of air showers triggered by the gamma photons. In addition, we must not forget that also the protons and other hadrons of the cosmic radiation produce Cherenkov light. Furthermore, the camera of MAGIC reacts sensitively to the effect of the atmospheric muons and many background sources of light (from the moon to passing cars). Moreover, it should not be forgotten that not all the photomultipliers of the camera always work uniformly, nor the fact that the atmospheric conditions can affect the quality of the observations, and the fact that the electronic components of the telescopes include a certain amount of noise into the signal etc.

That is, to detect the phenomena of cosmic γ-rays at all, the first task is to analyze the raw data carefully: the contribution of the background light must be excluded from the data; the hadron-induced events in the camera must be detected and excluded, too; the effects of the irregularities of the various components of MAGIC (mirrors, PMTs, etc.) must be corrected, etc.

But all this raises the question, whether such sometimes drastic changes (corrections, adjustments, clearing etc.) in the original data cause a disguised replacement of the empirical basis by a construction, which is to serve certain theoretical or practical interests of the physicists. Can we after such methods of analysis still speak about «discovery» of the phenomena? Would not rather «construction» or even «invention» be more suitable terms to describe the end products of the data analysis? In this section we will try to answer these questions.

For this purpose, it will be useful to divide the section into four subsections: in the first two subsections the two different data analysis chains are to be outlined, which are used by the members of the MAGIC Collaboration; in the third sub-section, the Monte-Carlo Simulations are discussed, which play a key role in the analysis of the MAGIC Data; and in the last subsection we will try to get some evidence for our discussion in the previous reports.

4.1 The Würzburg-Analysis chain

As noted in the second section of the chapter, the different teams within the MAGIC Collaboration have the advantage of great independence from each other. Therefore, it is not surprising that different methods for the analysis of the raw data of MAGIC has been developed and tested by different groups. Currently, two analysis chains are mainly used: the Würzburg and the Barcelona

Chain. Most teams use the Barcelona Method . But the Würzburg Chain may be a bit easier to describe, and so we will deal with this method first. It should be noted, however, that the main steps of the two methods are identical. These operations are as follows:

(1) Calibration of data	The response of the PMTs of the camera is considered here, and potential hardware errors are corrected.
(2) Background cleanup	At this point, the night sky background light is removed.
(3) Parameterization of the events	At this point, the recorded events are characterized.
(4) Gamma/Hadron separation	Here the hadron events are removed.
(5) Determination of the gamma excess from the source	At this point, the gamma background events are subtracted.
(6) Energy determination of the gamma events	The energies of the detected gamma photons are reconstructed at this point.
(7) Determination of the spectrum of the source	The differential energy spectrum of the source is calculated at this point.
(8) Production of a light curve of the source	The time variability of the source is determined at this point.

The way with which some of the steps of data processing are implemented is, however, different. In the Würzburg Analysis Chain, they are implemented by using the following computer programs:

1. Step of Würzburg Chain: CALLISTO

Callisto is the first program of the analysis chain. The goal of this program is to determine the number of photoelectrons that are released in the various events of a data set in each pixel of the camera. Since the PMTs and the other components of the camera do not always work uniformly, the recorded data must be corrected accordingly first. The program calculates conversion factors for the data of each PMT. In order to compute these factors, Callisto uses the data supplied by the calibration system – that is, the result of the transmission of light pulses of different intensity for the calibration of the PMTs of the camera [the so-called «calibration run»] – that are gathered before each observation with the telescopes.

Also the data of the so-called «pedestal run» –a number of 1000 randomly-triggered events that are recorded at the selected source position closer to the normal triggered data for the calculation of the camera noise– are used at this point. With the help of the calibration and pedestal data, Callisto determines the response of the photomultipliers of the camera in the actual data collection. The program can also detect defective pixels, and to some extent replace the missing data by interpolating the neighboring pixels.

2. Step of Würzburg Chain: STAR

The raw data corrected by Callisto are further processed using the STAR Program. At this point, it is important to clean up the recorded particle-air-shower events from the night sky background light, and to characterize the settled events. To this end, the program initially intends to delete the information of all pixels that contain no photoelectron induced by the air showers. This cleanup is done in two steps: in the first step all pixels are selected, which contain a signal of more than 8 photoelectrons (or 10 or 6 photoelectrons, depending on the severity of the selected «cleaning level»). These are then interpreted as «core» pixels. In the second step all the pixels that have at least a «core» pixel as neighbour and a signal of at least 4.5 photoelectrons (or 5 or 3) are selected All other pixels and also the «core» pixels that have no neighbouring pixels with a signal of at least 4.5 (or 5 or 3) photoelectrons are then deleted. The choice of these thresholds is based on Monte Carlo Simulations which have shown that the equilibrium between a strong cleaning with large signal loss and a weak cleaning with substantial remaining background noise is best reached with the mentioned numbers. The results of such a clearing are shower pictures like the following[184]:

---

[184] BRETZ (2006), 41. The photo shows two air shower events before and after the cleanup. Courtesy of the author.

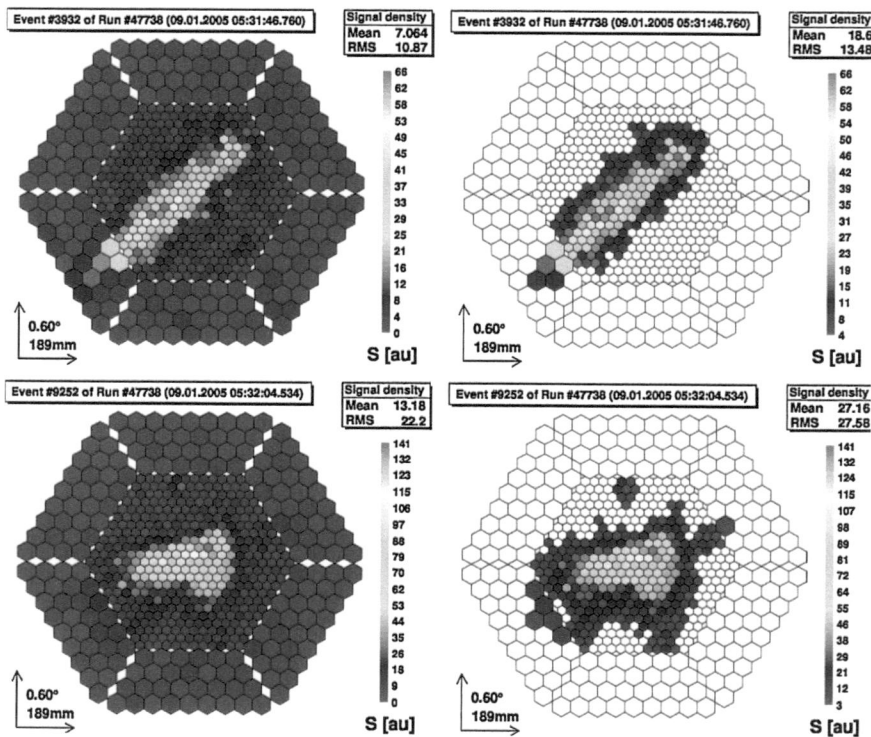

Of course, the cleanup of the background noise is more complete, the higher the thresholds for the maintenance of the pixels are. But the disadvantage of high thresholds is that many low-energy air shower events are lost. To increase the sensitivity of the method, there is the possibility to include the evolution of events in the «cleaning» process. In this variant, first a cleanup of the pixels with low thresholds (e.g. 6 and 3 photoelectrons) is performed. Then, in order to remove as many of the remaining background noise events as possible, a time adjustment is carried out: the core pixels whose signal does not lie in an interval of 4.5 ns with respect to the average time of an event are deleted. Likewise the other pixels are deleted, whose signal does not lie in an interval of 1.5 ns with respect to a neighbouring core pixel. The choice of these time thresholds relies again on studies of Monte Carlo Simulations.

After the clearing of the shower events, STAR implements also the characterization of the different events. This characterization is carried out with the help of a number of image parameters. Hillas introduced the most important parameters in 1985. They are, therefore, called «Hillas Parameters». The following picture shows some the Hillas Parameter for the characterisation of an event:

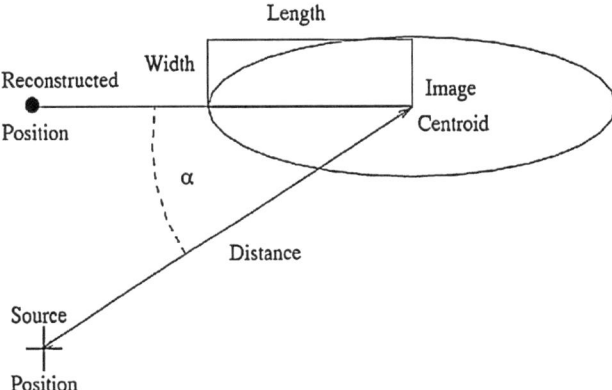

There are other parameters not shown in this image, such as «island» I which indicates the number of pixel islands of an event, or «area» A which represent the area of the event(i.e., $A = \pi \cdot Width \cdot Length$), or «size» which specifies the number of photoelectrons of a picture, etc.

3. Step of Würzburg Chain: GANYMED

After the recorded events have been characterized by various parameters, the third program of the analysis chain follows: GANYMED. The goal of this program is to retain as many gamma events as possible by discarding as many events of other types as possible. At this point in the process the crucial step of the signal/background separation, thus, takes place. It has already been noted that the ratio signal/background in this observation technique is very unfavourable (1/1000 to 1/10000). However, it is possible to extract the desired signals from the background noise. Because the by far most outweighing events in this background are hadron events –i.e. Cherenkov Flashes associated with the spreading of the particle air showers caused by the charged cosmic radiation–, and the shape of the camera images of such hadron events is usually different from the shape of the gamma events. This is because the hadronic air showers are clearly wider than the gamma-induced air showers, leading to wider event images, which often contain multiple «islands». Another kind of background are the muon events. But these events have a characteristic (arc or ring) shape which is clearly distinct from the shape of the gamma images:

The different forms of the shower pictures make it possible to reject all events which do not possess the expected form of the gamma events by restricting the permissible values of the different parameters. These restrictions or cuts in the space of possible parameter values are called «cuality cuts»[185]. The first action that is implemented by GANYMED, hence, is the removal of all events whose parameters do not meet the requirements of the «cuality cuts». Some «cuality cuts» are intended to remove insignificant events (like events that are caused e.g. by passing cars or are due to errors in the detector). Some of these cuts are for example the following:

- The number of pixels after cleaning must be higher than 5: N > 5
- The number of islands in the image must be less than 3: I < 3
- The proportion between the parameters «length» and «size» must be as follows:

$$LENGTH > -5(\lg(SIZE) - 4.7)^2 + 50$$

-etc.

In addition, there are «cuality cuts» that have been specifically conceived for the removal of hadron events and muon rings. The most important cut is:

$$AREA < c_2 \left(1 - c_4 (\lg(SIZE) - c_3)^2\right)$$

where c_2, c_3 and c_4 represent free parameters whose values can be determined by using Monte Carlo simulations.

In addition to the extraction of the gamma events, GANYMED computes the significance with which the observed source has been detected. To this end, the number of gamma events that are coming from the direction of the source are compared with the number of events classified as gamma events that are detected from an adjacent source-free direction[186].

---

[185] The use of «quality cuts» to select data is one of the most frequent steps in the processes of data analysis in astroparticles physics, and not only there but in all current physical investigations being carried out in circumstances where there is a strong background to be removed. However, there is not a «trouble-free» procedure, because the cuts can help to create «artifacts» or to make invisible real phenomena. For this reason, we give below some indications of the way in which the «quality cuts» are selected in the processes of analysis that we are detailing here. For a discussion of more details about these cuts, in relation to other branches of physics, we recommend the book: Franklin (2002), *Selectivity and discord*. In this book Franklin discusses several episodes in which selection criteria, or cuts, were used, sometimes improperly, and what safeguards exist in the practice of science against such procedures.

[186] For details about such calculations see e.g.: LI - MA (1983) 317.

4. Step of Würzburg Chain: SPONDE

After the gamma events have been extracted, the spectrum and the light curve of the source can be finally computed. To this end; however; the energy of each event must be determined. This is the task of the SPONDE program. For this purpose it compares the «size» parameter of the individual gamma events with the «size» distribution of a Monte Carlo Simulation. Since the connection between «size» and energy in the simulated events is known, the energies of the observed gamma events are identified indirectly by this comparison. However, the efficiency of this procedure depends crucially of the accuracy on the used simulations:

«It is, therefore, of utmost importance to ensure that the measured data can be described perfectly by Monte Carlo Simulations. For the more accurately the measured data is described by such a Monte Carlo Simulation, the more accurate and reliable physical results stand at the end of a data-analysis»[187].

Since the Monte-Carlo Simulations play such an important role in data analysis, we will consider them in more detail in the third subsection. But first, the Barcelona version of the MAGIC Data Analysis Chain is to be outlined.

4.2 The Barcelona Analysis Chain

The Barcelona Analysis Chain is the method of data analysis used by most groups of the MAGIC Collaboration, and can, therefore, be called the current «standard analysis chain» of MAGIC. As mentioned previously, the Barcelona-Analysis follows the same steps as the Würzburg-Analysis: calibration of the data, background rejection, parameterization of the events, gamma/hadron separation, determination of the gamma excess from the source, energy determination of the gamma events, and determination of the spectrum and the light curve of the source.

However, some of these steps in the Barcelona-Chain are implemented differently than in the Würzburg-Chain. In this subsection we will pay attention on these differences. But since other aspects of the procedure are very similar to those already described, we can leave many details aside. The Barcelona Analysis Chain consists of the following computer programs:

1. Step of Barcelona-Chain: CALLISTO
CALLISTO is the first program of the analysis chain. The goal of this program, as with the Würzburg-Chain, is to determine the number of photoelectrons which were released at the different events of a data record in each pixel of the camera.

---

[187] THOM (2009) 37.

It is not the same program as in the Würzburg-Chain, but there is no significant substantial difference between the two. Therefore, no further explanations are necessary at this point.

2. Step of Barcelona-Chain: STAR

This step also is similar to the Würzburg Chain. The night sky background light is rejected as described above, and the relevant events are characterized using the Hillas Parameters and other parameters.

3. Step of Barcelona-Chain: OSTERIA

After the cleanup and characterization of the events, the next step is the separation of the gamma signal from hadronic background. The implementation of this important step differs from the Würzburg-Chain. In the Barcelona-Chain the separation is implemented with the help of the so-called «Random Forest» method, which is implemented in two programs. The principle of this method can be explained as follows: first, a parameter space is defined using parameters such as «length», «width», «size», «area» that are (by experience) the most appropriate for the gamma/hadron separation. Each event, therefore, is characterized by a vector v in this parameter space. With the help of the select parameters a set of binary decision trees is derived then. The decisions that form the branches of the trees are of the following kind: «if "length" is larger than X, then ...», «if "size" is larger than Y, then ...» etc., whereby the order of these decisions is different from tree to tree. Then the vectors of the various events are classified by these trees[188]:

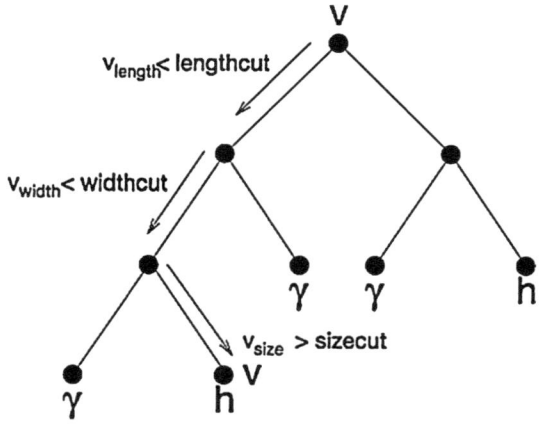

---

[188] In ERRANDO TRIAS (2009) 95. Original source: http://arxiv.org/abs/0709.3719. Courtesy of Thomas Hengstebeck.

In general, the classification of an event as a gamma or a hadron at the end of such a tree is not the same every time. Therefore, one calculates the average classification of each event. This parameter is called «hadroness» and with its help, the separation between gamma events and hadron events are carried out.
But in order to make an accurate gamma/hadron separation with the help of this method, the «cuts» of the decision trees must be optimized. This step is accomplished in the OSTERIA program by using Monte Carlo Simulations. That is, different decision matrices are tested by simulations, and those that provide the best results are selected for the analysis. To this end, it is important that the simulated data sets are similar to the real data sets. Therefore, it is common practice that the simulated gamma events are added to the real hadronic background of the measurements.

4. Step of Barcelona-Chain: MELIBEA

In MELIBEA the trained «random forest» matrices are used to compute the «hadroness» of the events, which had been previously selected and characterized in STAR.

5. Step of Barcelona-Chain: FluxLC

In this step, first the significance of the detection is calculated similar to the Würzburg-Chain. And then the spectrum and light curve of the source are produced. To this end, the «random forest» method is used again: a new collection of Monte Carlo Simulations goes through the whole analysis chain up to MELIBEA. With the help of these simulated data «random forest» matrices are then trained for the energy reconstruction. Finally, these matrices are applied to real data, and the searched spectrum and light curve of the source are computed.

6. Step of t Barcelona-Chain: CELESTINA

Finally, with the help of this program a sky map of the examined region is provided, in which the starting point of all reconstructed events classified as gamma events is specified.

4.3 Monte-Carlo Simulations

In the preceding subsections, we have seen that Monte Carlo Simulations play a crucial role in different places of the analysis chain. Therefore, this overview of the procedures for the analysis of the MAGIC Data would be incomplete if we did not include at least a short presentation of the simulations used in these procedures.

So what are the Monte Carlo Simulations? What is their origin? For what are they commonly used? And why do they have to be used here? What is simulated concretely? And which steps do these simulations contain?

Let's try to answer these questions successively – that is, starting with the question, what are the Monte-Carlo Simulations–:

> «Monte Carlo Simulations were used for the first time in the 1940s in connection with the construction of the nuclear bomb. Here the intention was to predict theoretically the interaction of neutrons with matter. These simulation-methods owe their name to the well-known gambling place Monte-Carlo. The basic idea of these methods is the use of random numbers that can be generated, for example, with a roulette wheel. Based on this central idea, a whole set of fundamental Monte Carlo Simulations had already been developed at that time. Today, these simulations rank among the most important numerical methods applied to many scientific, technical and medical problems with great success. Also in physics Monte-Carlo Simulations have been used for a long time for the energy calculation e.g. in the case of interactions and decays of elementary particles in the atmosphere or in the accelerator experiments. [In the MAGIC Analysis Chain] γ-events are simulated with the help of Monte Carlo-based algorithms, with which the origin of the shower pictures can be determined. One can recursively infer the energy of the primary particle, which thus enables an estimation of the energy of the primary particles in the Imaging Air Cherenkov Technique»[189].

The Monte Carlo Simulations, therefore, are about mimicking certain processes where with a certain probability different alternative developments can take place with a certain probability. A random number generator is used at the nodes of the process to decide which path is continued. This way, the end result is the description of one of the possible developments of the process investigated. If one repeats the simulation, then the final result will look differently due to the decisions made by the random number generator. That is, if one performs a series of such simulations, an overview of the expected results of the examined process can certainly be obtained.

The application of such simulations is inevitable in the context of the air Cherenkov telescopes, because there are no artificial gamma emitters, with whose assistance an absolute calibration of the detectors would be feasible. Therefore, the energy of the studied gamma photons can only be determined by comparing real and simulated events.

In order to simulate gamma events in Cherenkov telescopes such as MAGIC, the first thing necessary is to reproduce the development of the particle showers caused by the gamma photons in the Earth's atmosphere. Beyond that, the response of the different components of the telescopes to these hypothetical particle showers must be calculated. In the case of MAGIC these steps are

---

[189] THOM (2009) 40.

performed by three computer programs: CORSIKA, REFLECTOR and CAMERA. Let us briefly summarize how these programs work:

4.3.1 CORSIKA[190]

The original version of this program has not specifically been developed for MAGIC. The CORSIKA Program has been designed at the *Forschungszentrum Karlsruhe* for all ground experiments in astrophysics, for which it is important to simulate how the diverse particles of cosmic radiation interact with the atmosphere. It is, therefore, a very versatile simulation program that gives the user many options concerning the type and energy of the incoming particle, concerning the models of the interactions between the particles etc. This original software has been modified for the use in the MAGIC Analysis Chain: in this version the program determinates the interaction models of the particles used. Furthermore, the software was designed in such a way that, apart from the results of the interactions, further information about the incoming particles, etc. is retained. The version of the CORSIKA Program used for the analysis of MAGIC-Data is also known as MAGIC Monte Carlo Software (MMCS)[191].

The input of CORSIKA Program, modified for MAGIC, consists of input parameters such as the particle type, the energy of the incoming particles, the position and trajectory data (zenith angle and azimuth), the model of the atmosphere to be used and the number of events that must be generated.

On the basis of these parameters the program computes the particle reactions, which can take place in the atmosphere. With a certain probability the diverse interactions can lead to certain results. A random number generator is used to decide on the outcome after each interaction. (This is the reason why these simulations are called «Monte-Carlo Simulations»).

The main output of the MMCS program is the information about the Cherenkov photons generated in the simulated showers.

4.3.2 REFLECTOR[192]

The REFLECTOR program fulfils two different tasks: it simulates both, the atmospheric absorption of the Cherenkov Photons on the way to the telescope and the reflection of the photons in the mirror of MAGIC. These tasks are accomplished by the following steps:

---

[190] For a more detailed presentation of this program, see the official website of CORSIKA: http://www-ik.fzk.de/corsika/
[191] See a more detailed presentation of this CORSIKA-Version in SOBCZYŃSKA (2002).
[192] See the details of the REFLECTOR-programm in MORALEJO (2003).

«–In the atmosphere, there are two possible interactions the Cherenkov photons can carry out, which are both simulated within the Reflector program: Rayleigh Scattering from the air's molecules and Mie Scattering from aerosols. The absorption caused by both of these effects is simulated for the Cherenkov photons from the shower.

– The next step is to check if the incoming photon hits the mirror dish. If this does not, it is neglected further on.

– For all photons which hit the telescope dish, the absorption of the aluminium covering of the mirrors is taken into account.

– Then the mirror element which has been hit is determined.

– The reflection itself is simulated.

– It is checked if the reflected photon hits the camera.

– The arrival time for the incoming photons is calculated»[193].

For most of the steps, certain outcomes can occur with a certain probability. Again a random number generator is used to decide which result is accepted in each case.

The output of the program is information about the Cherenkov Photons reaching the camera of the telescope.

4.3.3 CAMERA[194]

Based on the output of the REFLECTOR Program, the CAMERA Program finally simulates the response of the PMTs of the camera, as well as the behavior of the trigger and the data acquisition system. In the end, data sets are produced that have the same form as the real MAGIC Data. Therefore, they can be analyzed in the same way as the real data with the help of the programs outlined in the previous subsections, and they can be compared with the real data.

Finally, it should be noted that most models which are used for the simulations outlined here are quite unproblematic –the model of the absorption of photons in the mirror, the reflection model of the photons, the model of the response of the PMTs, etc. Because these models are based on sound knowledge of optics, electronics, etc., and beyond that for modelling the individual components, test results are used, that were revealed by the checks of the various elements of the telescopes. Most problematic is the simulation of particle showers in the CORSIKA Program, because much information about the cross-

---

[193] DOERT (2009) 20.
[194] See the details of the CAMERA-program in BLANCH (2001).

sections of the different processes is unknown and must be extrapolated from the well known data from particle accelerators. But these difficulties rather concern other experiments in astroparticle physics than MAGIC (or more generally than the Cherenkov telescopes). For the most unknown factors required in CORSIKA are connected with the hadronic and not with the electromagnetic showers produced by the gamma photons.

4.4   Philosophical Comments about the Admissibility of the Analysis Method of the MAGIC Data

Following the brief presentation of the data analysis chains used by the members of the MAGIC Collaboration, we can now discuss the question of whether the data analysis is the stage of the research process in which a construction of the phenomena motivated by theoretical (or other) interests can take place. Assuming that we would be interested in the confirmation of a specific effect (e.g. a certain energy dependence in the arrival time of the gamma photons from GRBs), could we bias the data analysis so that the desired effect is «discovered»?

At first glance, it seems possible to influence the results of the data analysis mainly in two places: first, in the selection of cuts (e.g. trigger cuts at the beginning of the process or the parameter cuts for the gamma/hadron separation) and secondly, in the selection of the theoretical models of the processes on which Monte Carlo Simulations are based.

But if one looks closely at these places, it is very difficult to see at what point there really are any degrees of freedom to arbitrarily construct the desired effects. Because the cuts are not arbitrarily selected. Instead, one always uses the cuts, with whose assistance the best results were obtained in the analysis of data sets from Monte Carlo Simulations.

And with regard to the simulations, REFLECTOR and CAMERA are largely unproblematic, because the response of the mirror elements and the various components of the camera –simulated in these programs– is well known due to the tests accomplished in the development of the MAGIC Telescopes. The most difficult point of the simulation chain in MAGIC is undoubtedly the CORSIKA Program, because in this program almost an infinite number of interactions can be taken into account, many of which are only known incompletely so far. That is, at this point an estimation of the most relevant interactions is required, which certainly is accompanied by the risk of possible bias in the results.

But precisely for this reason the focus in the development of the program has been primarily on the inclusion of more and more particle interactions in

CORSIKA[195] in the last thirteen years, with the goal of t finally taking into account all relevant factors in the models of the simulated particle showers. This development is still ongoing, but now the trend has already become apparent that the inclusion of new aspects in the current models only leads to slight deviations of the final results[196]. In addition, CORSIKA is developed primarily[197] by a research group which neither belongs to the MAGIC Collaboration nor has created the program exclusively for the Cherenkov telescopes. Facts that certainly ensures the independence of the program with regard to all sorts of particular goals of this collaboration.

But if the selection of cuts are based on the results of the analysis of simulated data and the simulation programs were developed without regard to the members of the collaboration doing the data analysis, and if these programs now provide such a solid description of the processes investigated that the new upgrades only cause secondary changes in the results, then it is not easy to see, how the phenomena could still be constructed under such restrictions.

So much about the question of whether it is plausible that the data analysis represents the point where a deliberate construction of the phenomena investigated by MAGIC might be motivated by particular interests. But could it not be the case that the procedures of data analysis cause an unintended construction of the examined phenomena to take place? In other words: could it not be possible that unplanned artefacts are created in the process of data analysis with the effect that a realistic view of the research with the MAGIC telescopes is out of place? And could it not even be the case that such artefacts developed unintentionally, finally serve as useful elements for the proof of any appreciated hypotheses? In this case one would have to admit that also without a conscious manipulation of the data analysis this procedure is to be understood in the sense of constructivism.

However, several facts speak against this possibility, only two of which will be mentioned here (for the sake shortness): (1) The good compliance of the results that are achieved by the application of the two alternative analysis chains

---

[195] See e.g. the updates of CORSIKA listed in the following website: http://www-ik.fzk.de/~heck/publications/

[196] An important example is the update of the CORSIKA- program fort he simulation of the effects of charm interactions for the simulation of the development of hadron showers. See for this: http://www-ik.fzk.de/~heck/publications/fzka7366.pdf
The result of this update ist he following:
«–The inclusion of explicit charm production in simulations gives no significant differences for most observable shower parameters.
–As expected, the number of 'direct' muons with energies above 100 TeV is slightly increased by the charm production».

[197] That is, if the adjustment of the program CORSIKA to the concrete conditions of the MAGIC observations in the MMCS is not considered.

of MAGIC, and (2) the congruence of the results of MAGIC with the results obtained with other instruments of astroparticle physics in the so-called «joint campaigns».

Here are some explanations:

To (1): In the previous subsections we have noted the clear differences between the Würzburg and the Barcelona analysis chains. If any unintended artefacts should be generated in the process of data analysis, then one should expect (due to these differences) that the results of both analysis chains would differ significantly from each other. For each method will basically generate different artefacts. However, the parallel analyses of different data records implemented so far rather show a good compliance of the results of both chains[198].

To (2): There are a number of celestial bodies – particularly some active galaxies – which have been objects of so-called «multiwavelength campaigns» or «joint campaigns» for years. Typically, the objects are simultaneously observed in these campaigns by radio telescopes, X-ray satellites and two or three Air Cherenkov telescopes. It should be noted that the different large Cherenkov telescopes are operated by independent collaborations that have developed different methods of data analysis.

If any unintended artifacts should be generated in the process of data analysis, then one should expect (due to the differences in the methods of data analysis applied to the data sets of the different Cherenkov telescopes and the still larger differences, for instance, between the gamma and the X-ray detectors) that the results of each collaboration strongly differ from each other. Instead, however, the fact is that the gamma flares of the AGNs can be observed by all Cherenkov telescopes involved in the campaigns. And there often also is a correlation between the gamma and X-ray flares[199].

All these facts indicate that the results of the MAGIC Data Analysis represent no artefacts (and generally, no constructions), but they rather are to be interpreted as a description of real phenomena. Thus, we are now able to draw some conclusions for our debate between realism and constructivism. This is the goal of the next section.

5. Some Lessons to be drawn from MAGIC regarding the Debate between Realism and Constructivism

Imagine a theoretical physicist who has been turned into an inveterate constructivist by reading the works of Collins, Latour, Pickering, etc. Our physicist does

---

[198] To this point see e.g. THOM (2009) 70-73.
[199] Some recent examples can be found e.g. in MAGIC COLLABORATION (2009) section: «Multiwavelength and joint campaigns».

research based on a certain theoretical school which deals, for example, with the problem of dark matter. As he is an intelligent physicist he has created an elegant model of dark matter after years of hard work, which possesses a number of interesting theoretical features.

If our physicists would be a realist he surely tried to promote his model with the goal that other experimental physicists do the necessary observations in order to examine whether there are some natural phenomena which can be derived from his model. But our physicist is not a realist but a constructivist. Therefore, he thinks that these phenomena are not to be discovered, but constructed. Fortunately, it involves phenomena of the high-energy cosmic gamma radiation.

This situation is very favourable from a constructivist point of view, because it is a new research area, with new tools for the observation of new phenomena are currently being developed. And in accordance with the typical constructivist views these are ideal circumstances for the construction of phenomena.

Our good scientists, thus, contacts a team from the MAGIC Collaboration, asking, for example, to prove the existence of a certain number of gamma photons, originating from the halo of the nearby dwarf galaxies, with the help of the new (and still in the commissioning phase) telescope MAGIC II. What answer could he get? The members of the research team would probably listen to such a proposal probably respectfully and attentively. But they would also point out that some observations of dwarf galaxies have already been carried out with MAGIC, without any significant gamma flux having been detected so far[200].

But such a response could, however, barely puts off an inveterate constructivist from his resolutions. He would friendly respond that the telescopes are to be modified in such a way that the new data become consistent with the desired gamma flux. Or alternatively, the data analyses are to be controlled in such a way that enough gamma photons are detected from the halo of the galaxies.

I cannot say whether the members of the experimental-physic team would still be friendly towards our theoretical physicists. But if they still are they would certainly call his attention to some points discussed in the 3rd and 4th section of this chapter:

With regard to the hardware, the components of MAGIC are only used after a careful examination, and they cannot be arbitrarily replaced or arranged differently; the whole telescope was calibrated due to observations of the Crab Nebula and other known sources; and even with the development of the first air Cherenkov telescope (WHIPPLE) which significantly proved some gamma sources, left little room for free design , because the results of the observations underwent a number of consistency checks. And with regard to the software for data analysis, it consists of programs that leave hardly any design flexibility, and

---

[200] See e. g. ALIU, E. et al. (2008).

which cannot be randomly replaced by other programs because the analyzed data must be consistent with the results of other methods of analysis and other detectors.

One would make our physicist aware of the fact that the gamma photons from the halo of the dwarf galaxies, as suggested by him, do exist or do not exist. If there really is a certain number of such photons, it should be possible to detect them with the MAGIC Telescopes –the way they actually are!–. If nothing is detected, it is likely that the theoretical hypothesis is wrong, and the underlying model, as beautiful as it may be from a theoretical point of view, does not fit the real mode of being of nature.

That is, as far as the way of working in the MAGIC Collaboration represents a typical example of the practice within natural sciences –and it, indeed, is a typical example, at least with regard to physics–, the realistic interpretation of the development of the research seems to be more reasonable than the constructivist interpretation.

But if this is the case, why does Pickering's constructivist argument in relation to the discovery (or construction?) of the weak neutral currents *prima facie* seems to be so believable? In my opinion, this is due to a feature of the example examined by Pickering, namely, the fact that the discovery of neutral currents was generally accepted, although only two collaborations in the world observed the desired effect with the help unique detectors. This behavior on the part of the scientific community can be explained by the particular circumstances of research in particle physics. For in this field there are very few (often only one) particle accelerators which can be used for the observation of a particular phenomenon. One, therefore, tries to compensate the absence of tests carried out by independent teams and with other instruments by applicating a strict procedure of data analysis and by increasing the number of tests of the results within the collaboration responsible for the observation. And in my opinion, the study carried out by Galison regarding the discovery of weak neutral currents shows that this procedure can be very reasonable, and that the results of such a practice can be reliable in spite of everything.

But if one does not look on the details of the data analysis of such experiments, and only regards the process globally, the question will certainly arise, whether somewhere in the complicated process of building the detectors and data analysis, certain decisions were made that led the results in a certain direction. I think this is the reason why the example provided by Pickering as a hypothesis makes it seem plausible that a construction of the desired phenomena has taken place.

In most areas of physics –including astroparticle physics– the situation, however, often is quite different: there are different collaborations which can independently examine the announced new phenomena (with different kinds of detectors). The announcement of e.g. a particularly strong gamma flare from an

AGN can be controlled by different Cherenkov observatories. Such flares may be detectable for the gamma satellite, and usually for the X-ray satellites, too. Under these conditions the idea of a construction of such phenomena seems bizarre.

What we can learn from this is, that the ideal field for the possible occurrence of constructivist dynamics of research is a major collaboration that works on the construction of a new and very expensive, complex detector with the help of which the detection of a new set of phenomena is expected. A result that could not be examined by other collaborations now and in the near future

The funds invested into a collaboration of this kind will cause a great pressure to succeed. The prospects of fame and the chance of providing new jobs for members of the collaboration will increase the pressure even more. Whereas the fact that there are no other groups that can critically analyze the results, probably induces the aforementioned results to be interpreted without all suitable precautions. Is it at all possible to do research in a realistic rather than constructivist manner when such circumstances occur?

We will try to answer this question in the next chapter based on the example of the IceCube Collaboration.

4. Chapter: IceCube

1. Introduction

In the last chapter we gave an overview of the instruments and methods of analysis used in the MAGIC Collaboration. The consideration of the individual points was guided by the question of where the hardware or the software analysis of MAGIC could offer opportunities to construct desired phenomena from raw data. We found that there is no plausible place for such an operation. It was also pointed out that the results of MAGIC are always subject to strict control by other collaborations that use other devices. That is, at least at this point the hypothetical artefacts and constructions that could result from the practice of the MAGIC Collaboration, would have virtually no chance to be established as accepted phenomena.

But what would happen if this last control level was missing? For this is the case when a scientific collaboration has detectors and other technical resources that are far superior to the resources of all other research teams in certain fields. We have to address this issue seriously, because the situation indicated is already occurring in certain border areas of physics and will continue to do so probably even more often. (Which is no bold with regard to size and complexity of the devices that are currently used, for instance, in particle physics or in astrophysics).

Would factors such as, for example, the pressure to succeed (to justify the project costs) or the desire of scientists to make a career and acquire fame, in this case not necessarily lead to a dynamic of the research in the sense of social constructivism?

In this chapter we will examine this question using the IceCube detector –still under construction– and the work in the IceCube collaboration as an example. To this end the chapter is divided into the following sections:

First, [2nd section] some of the social aspects are outlined that can affect the work within a large collaboration. Then we will try to find out whether such factors can explain the dynamics of research in the IceCube collaboration in the rest of the chapter. In the following [3th section] the IceCube collaboration is presented. Then [4th section] a description of the IceCube detector follows together with a summary of the procedure of data analysis used by the members of the IceCube collaboration. In this section the AMANDA experiment is briefly outlined too. (The AMANDA is the predecessor experiment of IceCube, and was integrated as part of the new research facility). Some examples of the technical development of the components of the IceCube Facility are also mentioned, although they are not discussed in such detail as the examples in the last chapter. And finally, [5th Section] some evidence for the controversy between realism and constructivism are to be gained from the research of the IceCube Collaboration.

2. «Social Forces»: Extra Scientific Aspects of Large Scientific Collaborations

Looking back on the arguments of scientists who participated in the discussions of the «science wars» in the nineties, Bruno Latour, one of the leading representatives of the social constructivist approach, wrote the following ironic remark:

> «[...] scientists made us realize that there was not the slightest chance that the type of social forces we use as a cause could have objective facts as their effects»[201].

The remark is ironic insofar as Latour –according to his constructivist perspective– does not believe that natural sciences describe such things as «objective facts». But I didn't mentioned this quote just as a source of amusement but because it very clearly shows the fundamental difference between the realist and the constructivist view on the dynamics of scientific progress: while realists see the dynamics of the research aiming to discover and explain «objective facts», in order to create a picture of the real structure of the world, constructivists essentially regard dynamics as an interaction of «social forces», from which no objective facts –if there are any – can be derived .

When trying to understand the development of a particular line of research one should, according to the constructivists, not so much consider the content of the scientific publications which include the experimental and rational justifications for the acceptance or rejection of the theoretical hypotheses. One should rather consider the extra-scientific social forces (not usually mentioned in the scientific articles) such as career opportunities, the chances of receiving state funds, the academic trends and so on. For these are the real decisive forces in the process of research. Ian Hacking has expressed the same idea as follows:

> «Rationalists think that most science proceeds as it does in the light of good reasons produced by research. Some bodies of knowledge become stable because of the wealth of good theoretical and experimental reasons that can be adduced for them. Constructivists think that the reasons are not decisive for the course of science. [...]
>
> The constructionist holds that explanations for the stability of scientific belief involve, at least in part, elements that are external to the content of science.

---

[201] LATOUR (2005) 100.

These elements typically include social factors, interests, networks, or however they be described»[202].

In the previous chapters we dealt with dynamics of research in astroparticle physics almost exclusively from the internal perspective of the arguments discussed in scientific papers regarding the rejection or acceptance of a hypothesis. We talked about calibration methods, cross checks, tests for verifying the operation of measuring devices, etc. Is it, however, not conceivable that all these arguments represent only subsequent rationalizations of decisions based on other motives? Didn't we concentrate too much on these irrelevant rationalizations, and have neglected the crucial factors? In this chapter we will try to find out whether an «external» approach (by considering social factors) to the dynamics of the research enables us in a concrete case to understand the development of the research better than by considering the «official» discourse of the physicists participating in this development.

We want to analyze a case where the power of the social factors should be very significant, at least in principle. And so it is best to focus on the work within a large scientific collaboration. This might at first appear to be a very specific example, but it is not. And the reason is as follows:

In the course of history the limits of science and especially of physics have increasingly been moved forward into areas that can only be explored with the help of very complex facilities and measurement devices in cooperation with dozens or even hundreds of scientists. As from the second half of the 20th century experimental physics has therefore been becoming a science of large collaborations in many areas. That is, if it becomes apparent that the work within such collaborations is essentially influenced by extra-scientific social factors, this should be considered as a clear support for the constructivist perspective.

In fact, there are numerous authors that indicate the weight of the extra-scientific factors in the dynamics of large collaborations. Take, for example, the deliberations of the astrophysicist López Corredoira in connection with the research with the existing telescopes. It is a long quote, but it is worthwhile to be included in full length because this quote very well summarizes the problem of the influence of extra-scientific (social) factors in the current research:

«Before using large and even not so large telescopes, a tribunal of specialists must be convinced that something will be found which is already expected. In order to use these great installations, some forms must be filled up between six and twelve months before the observation date. In these forms, one must clarify what finding is anticipated as a result of the application of their measurements and observations. The tribunal of specialists must be shown the purpose of pursuing the observations. In addition, profile data on the re-

---

[202] HACKING (1999) 91-92.

searchers must be filled. Of course, the greater the history of observational publications and telescope time that a researcher has had in the past, the higher the probability of gaining further telescope time. Therefore, the researcher will be able to publish more papers than anybody else, although all the papers are similar and without any worthwhile or new ideas but prestige will be increased and enhanced. [...]

Nonetheless, the biggest problem for the advance of science is that new ideas are not welcome among the tribunals that authorize telescope time. If somebody applies for telescope time in order to test the predictions of an alternative theory, rather than the standard one, the proposal is most likely to be denied. We are not talking about amateurs for whom some crazy idea has occasionally gone to their heads; we are talking about great professionals whose only defect is in doubting the ideas which all the rest of scientific thinking considers untouchable. The system does not support an ideological plurality within the science. It is said that there is freedom in research, but this is just a lie as are so many other statements [...]. Of course, anyone can think what he wants to think, but the installations, the prestigious publications and the propaganda are only for those who want to make a science a reconfirmation and underscoring of certain prejudices, rather than an opening towards new horizons.

It might also happen that somebody presents an idea or an objective that is interesting for its study, but the work cannot be developed because the tribunal does not make telescope time available. [...]

The scientific body politics is convenient for large flows of capital. A telescope such as the 10-meter one that is being built in La Palma with a Spanish budget costs the huge amount of around one hundred million euros. If one looks for high amounts of money, space telescopes and the great satellite research projects costs go beyond one billion euros, and are paid for by several countries. Apart from the building costs of the telescopes and satellites, there is also the maintenance expense. In total, taking into account the average life of a large ground or space telescope, each observation hour costs thousands or tens of thousands euros. Therefore, it stands to reason that the use of the telescopes by the first barmy person who promises the moon and stars should be avoided at all costs. Moreover, nowadays, in order to obtain such huge amounts of money, it is better to show a solid image of science, an image that indicates science knows where it is going and has a clear view on problems that have been solved and those that are to be solved when money is available for research and instruments, etc. The image of a pluralistic science, entrenched in discussions about fundamentals, is not sufficiently interesting to attract investment money. Huge amounts of money are not invested for the purpose of allowing the scientist to play at guessing how nature is. Those sums, however, are invested in order to obtain a firm product far away from the speculative wordiness of philosophers. In other words, science is bought, and the one who pays has the right to demand the fruits are superior to those obtained for a lower price. All is accounted for with regard to the sums: the number of publications obtained with a telescope, the number of citations obtained by those papers—called "impact", which is a parameter as related to the quality of a work as is the number of people comprising an audience for various television programs. In the end, the reports must mention how profitable an investment has been. Parameters such as genius, creativity, mental lucidity and other human factors are not included in these reports.

Spontaneity has no place in actual science. Neither, is there a place for fortuitous discoveries favouring those who suspect that something in astrophysics is not following an appropriate path. Almost everything is planned in order according to programming and forecasts made many years in advance (it takes around fifteen years from the beginning of the plans to build a satellite until it is launched). Predictability was seldom present in the long history of science. Many times, science has had to walk back a way, to retrace its steps, before taking some new way or approach. Many surprising discoveries have been done by pure chance. However, contemporary system is apparently surer than the science of any previous time in that there are no historical errors; at least, if there are some errors, then the system tries to delay their discovery as long as possible. It is an apparent paradox that the greater the possibilities of science to observe and make experiments, the greater the obstacles—rather than motivation—for its advancement. Does astronomy move backwards with the advance of technology? There is no doubt that astronomy is an observational science that has larger 'possibilities' with better instruments. However, the control of science by the system is larger when larger telescopes are available, so private initiatives are blocked if they are in disagreement with established standpoints»[203].

The criticism of López Corredoira regarding the methods of current scientific collaborations seems to be plausible at first. And it can be understood as a confirmation of the constructivist position, although the author himself described the situation as an aberration and not as the essence of all scientific activity. However, the criticism remains very general and it would certainly be useful to substantiate e.g. the «predictability» of the development of research in the current system with the help of concrete examples. Is the science of the great collaborations really so predictable and so sealed off from surprising new developments, as López Corredoira states? We will discuss this question in the 5th section

---

[203] LOPEZ CORREDOIRA (2009) 166-168.

based on the example of the IceCube collaboration. But for this purpose it is necessary to outline the IceCube collaboration, and the IceCube experiment first. That is what the next sections are about.

3. The IceCube Collaboration

In the first paragraph of «governance document» of the IceCube Collaboration, the goals and instruments of this research partnership are briefly summarized as follows:

> «The IceCube Collaboration (the Collaboration) is an organization of scientists who collectively participate in a research program to study high-energy neutrinos from cosmic sources. The Collaboration uses the IceCube Observatory at the NSF South Pole Amundsen-Scott station for this research program. The IceCube Observatory consists of a surface array, IceTop, and a deep ice array IceCube. The IceCube array includes the strings of the former AMANDA detector.»[204]

The mentioned detectors (IceTop, IceCube and AMANDA) are described in the next section. But for the purposes of our study it may be useful to report about the composition and structure of the research community that uses these instruments first. In October 2009, the IceCube Collaboration consisted of over 250 members from more than 30 research institutions from Belgium, Germany, England, Japan, the Netherlands, Sweden and the USA[205]. The total funding for the work of the collaboration between 2004-2008 amounts to $ 271 million. Consequently, the IceCube collaboration is a large group of scientists who participate in a vast and complex experiment. To coordinate the work of individual researchers and teams –in order to make collaboration function– it is necessary to introduce some committees and leading positions. The most important are the following:

(1) The Collaboration Board
The collaboration board regulates both the organizational issues as well as all scientific and science-political activities of the collaboration. The board supervises the following matters:

«– science policy and goals

– membership

– data access

– publication

---

[204] Source: http://www.icecube.wisc.edu/collaboration/governance.php
[205] Source: http://www.icecube.wisc.edu/collaboration/authorlists/2009/10.html

– representation of IceCube at topical and general conferences
– analysis teams
– education and outreach»[206]

Each institution participating in the IceCube Project is represented on the board. The number of representatives of each institution depends on the number of its doctors.

(2) The Spokesperson
The spokesperson or speaker leads the collaboration board. He/she is a non-voting member of the collaboration board, which is elected by the doctors within the collaboration. His/her tasks are the following:

«– organizes and chairs Collaboration Board meetings

– during the IceCube construction phase is the interface between the collaboration Board and the Project Office, communicating with the Project Office on behalf of the Collaboration Board.

– arranges general Collaboration meetings

– speaks for the Collaboration in interaction with the scientific community

– speaks for the Collaboration in interaction with the general public

– selects members of Collaboration advisory committees subject to concurrence by Collaboration Board majority vote

– during the IceCube construction phase communicates with the International Oversight and Finance Group (defined in the Project Management Plan) on behalf of the Collaboration Board.

– calls for and oversees formal votes on particular issues»[207]

(3) The Executive Committee
In consultation with the collaboration board the speaker appoints (for 2 years) an executive committee. The task of the executive committee is to advise the speaker on the preparation of proposals which have then to be discussed and approved by the collaboration board. The executive committee also advises the speaker on his/her daily decisions.

(4) Other Committees
The speaker selects (with the approval of the board) members of the collaboration, who are requested to form committees for the coordination of the most important activities of the project. In particular, there are committees for: [1] the data reduction and data analysis; [2] the detector monitoring, maintenance and

---

[206] Source: http://www.icecube.wisc.edu/collaboration/governance.php
[207] Ibidem.

calibration; [3] the presentation of results as speaker of the whole collaboration at scientific meetings [Speakers Committee], [4] The publication of the collaboration results in scientific journals.

The diverse committees and working groups ensure a steady cooperation between the research teams. Additional information exchange between the members of the collaboration happens –as in the case of the MAGIC Collaboration (and common practice in other collaborations, too)– mainly through the meetings where researchers of the various teams have the opportunity to meet in person. Every year there are at least two major «collaboration meetings» with all members of collaboration being invited. The collaboration meetings are organized alternately by the different institutions participating in the collaboration. Within the scope these meetings a meeting of the collaboration board also takes place.

If one compares the structure of the IceCube collaboration with the structure of the MAGIC collaboration, outlined in the last chapter, it immediately becomes apparent that although the goals and instruments are different the organization of the teamwork basically is identical. In fact, the structure of the two collaborations is very typical. But it is precisely for this reason that the answer to the question, whether these circumstances allow at all a realistic interpretation of the results obtained by the research of IceCube, can justifiably be transferred to numerous other cases. Since this represents the main topic of the discussion in this chapter, we will close this section with some short remarks about the possible susceptibility of the described structure to constructivist developments. Let us consider the following question: does the form of the IceCube Collaboration favour the production of constructivist science?

Basically yes. For the supervision of the calibration of the detector, the data analysis and all other important steps of the research process as well as the authorization of the publications of the collaboration is ultimately led by a small group of scientists, also responsible for the petition and application of research funds.

Under these circumstances it is only natural to presume that the speaker, the executive committee and the members of the board are working under a significant «pressure to succeed». If one also takes into account that currently there are no further neutrino detectors the size of IceCube, which may provide for independent control of the results of this collaboration, the conclusions of López Corredoira cited in the previous section become very appealing and plausible:

> «Spontaneity has no place in actual science. Neither, is there a place for fortuitous discoveries favouring those who suspect that something in astrophysics is not following an appropriate path. Almost everything is planned in order according to programming and forecasts made many years in advance

[...]. Predictability was seldom present in the long history of science. [...] However, contemporary system is apparently surer than the science of any previous time in that there are no historical errors»

Are the future results and achievements of the IceCube Experiment pre-programmed right from the beginning? Of course, one can only speculate at this moment. But instead we can move the discussion a step forwards by examining two different points, i.e.: (1) the actual activities of the IceCube Collaboration; and (2) the relationship between expectations and results in the predecessor experiment of IceCube (AMANDA).

The first point is discussed in the next section, and the second one in the last. To conclude this section I would therefore like to pay attention only to one aspect that can be important for our discussion, and that we haven't considered yet: the realistic basic attitude of the collaboration members, which can be gathered from the different expositions of the goals of the project.

Let us regard, for example, the «IceCube Preliminary Design Document», which provides the basis for the whole project. If we read the scientific motivations for the construction of the IceCube Detector [section 3 of the document], we find that a number of models of the sources of cosmic radiation (as well as models of dark matter and exotic matter) are mentioned, which entail the production of high-energy neutrinos. The IceCube Experiment could play a key role in examining whether these models are correct. In other words, it is about checking whether certain models correctly describe the real nature and dynamics of the studied objects or whether they are to be considered false and be refuted. That is clearly a realistic attitude. These arguments reveal the curiosity of physicists to discover how the physical reality actually is.

In my opinion this realistic attitude and curiosity presents a kind of counterbalance to the extra-scientific motivations (such as the aforementioned economic and science-politically motivated pressure to succeed) which could result in a constructivist development of research. The question to be answered therefore is what factors dominate here in the end: what weighs more in the end? The will to achieve true knowledge of the physical reality or the social factors that suggest a predictable construction of the phenomena? We will try to answer this question in the following sections.

4. The Work of the IceCube Collaboration

4.1 Introducting Remarks

IceCube is a neutrino detector that is being built at the South Pole. The following image shows the exact place of this device[208]:

Some parts of the detector are already in operation and provide data that is analyzed in the collaboration. When the whole instrument is completed, IceCube will have a volume of about 1km^3, with 5160 optical modules for the detection of Cherenkov light. [An image of IceCube and some general remarks about this kind of detectors can be found in the subsection 3.6 of the second chapter of this book]. The passage of neutrinos through the detector certainly does not cause Cherenkov light, because only charged particles can polarize the ice molecules (or other media), and the neutrinos are obviously no charged particles. But it occasionally happens that a muon neutrino interacts weakly with the medium, and in this interaction a muon is produced. The trajectory and energy of these

---

[208] Source: «IceCube preliminary design document». In:
http://www.icecube.wisc.edu/icecube/static/reports/IceCubeDesignDoc.pdf p.11.

muons can be reconstructed by using Cherenkov light tracks. And from this information the energy and the direction of the initial neutrino can in turn be derived. IceCube, thus, is a detector specifically designed for the detection of high energy muons (and hence of muon neutrinos) in the ice.

In principle, one could use such an instrument for the detection of electron neutrinos, too. But the electrons lose their energy very quickly in the ice (as also in other media) due to *bremsstrahlung*. Therefore, one should place more optical modules in the ice to detect such interactions.

With regard to the tau neutrino – the third neutrino flavour – the main problem is that the tau leptons, which can result from the weak interaction of these neutrinos with the ice molecules, decay very fast. A fact that makes their detection significantly more difficult. On the other hand, the tau events possess special features that simplify the identification of such events. For both the production and the decay of tau leptons are connected with the production of particle showers, which are in principle detectable with the help of the optical modules. This would cause a so-called «double bang event» in the detector. However, such events can be observed in a $1km^3$ detector only at very high energies (from about 5 to 20 PeV) because of the rapid decay of tau leptons. This means that only a few dozens of such events are expected to be observed with IceCube per year.

Due to these reasons the design of the IceCube aims to optimize the detection of high energy muon neutrinos, while maintaining some sensitivity of the device for the other neutrino flavours.

What phenomena does one try to observe with IceCube? The «IceCube Preliminary Design Document» lists the following points of motivation for building the detector:

« A well-designed neutrino telescope can

– search for high energy neutrinos from transient sources like Gamma Ray Bursts (GRB) or Supernova bursts;

– search for steady and variable sources of high energy neutrinos, e.g. Active Galactic Nuclei (AGN) or Supernova Remnants (SNR);

– search for the source(s) of the cosmic-rays;

– search for Weakly Interacting Massive Particles (WIMPs) which may constitute dark matter;

– search for neutrinos from the decay of superheavy particles related to topological defects;

– search for magnetic monopoles and other exotic particles like strange quark matter;

– monitor our Galaxy for MeV neutrinos from supernova explosions and operate within the worldwide SNEWS triangulation network;
– search for unexpeted phenomena.»[209]

It must be noted, however, that not all points are equally important. It would, for example, not make sense to build IceCube mainly for the monitoring of supernova explosions in our galaxy, as detectable events of this kind only occur once in decades. One can also not justify the cost of the experiment alone with the hope to observe unexpected phenomena. Otherwise, the search for monopoles and other exotic particles is certainly interesting, but it would be too risky to invest funds in the order of several million Euro mainly for the search of these particles, since the existence of the exotic particles mentioned in the list is anything but certain (or even probable).

In fact, the search for high-energy neutrinos from GRBs and AGNs represents the main motivation for building the IceCube detector. For many typical models of such sources require a significant emission of neutrinos that are detectable with a 1km^3 detector. In the next subsections we will briefly describe how this search is taking place.

4.2 The IceCube Detector

The IceCube detector consists of three components: (1) the *InIce* detector; (2) the *DeepCore* area; and (3) the *IceTop* detector. These components can be described as follows:

(1) The InIce Detector
This part of IceCube consists of 80 strings placed in the ice at a depth of 1450 to 2450 meters. Each string has 60 optical modules for the detection of Cherenkov Light. The distance between the modules in each string is 17m. The various strings form a hexagonal structure, and the distance between a string and the six nearest neighbour strings is 125m. Until December 2009, 59 out of 80 strings were already placed and the corresponding modules were already providing data. The completion of the detector is planned for the season 2010/2011. The following figures show the planned string structure of IceCube and the placement of a string in a drilled hole in the ice[210]:

---

[209] «IceCube Preliminary Design Document». In: http://www.icecube.wisc.edu/icecube/static/reports/IceCubeDesignDoc.pdf p.12.
[210] Source of all following pictures: http://gallery.icecube.wisc.edu/ (reproduced with permission).

4. Chapter: IceCube 177

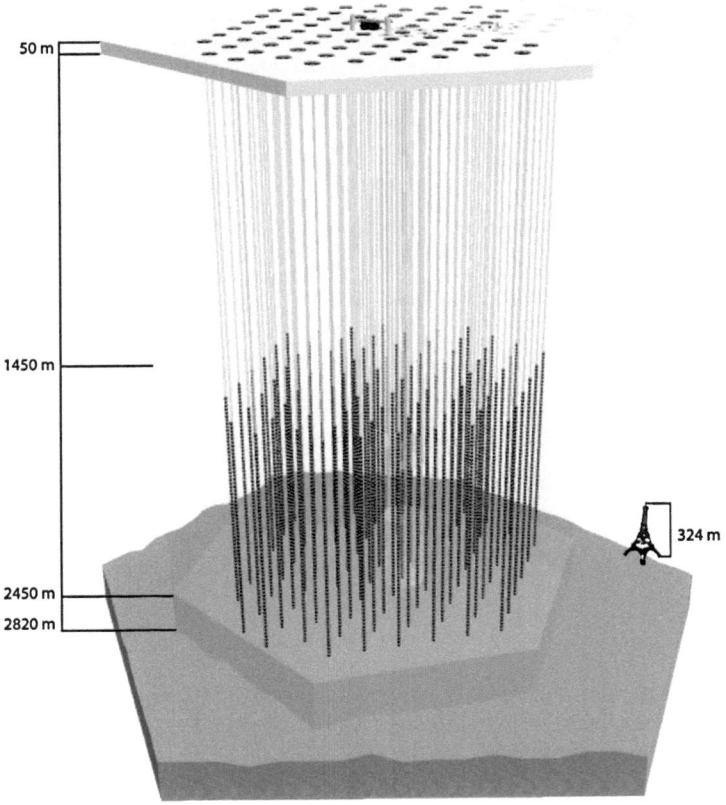

178 4. Chapter: IceCube

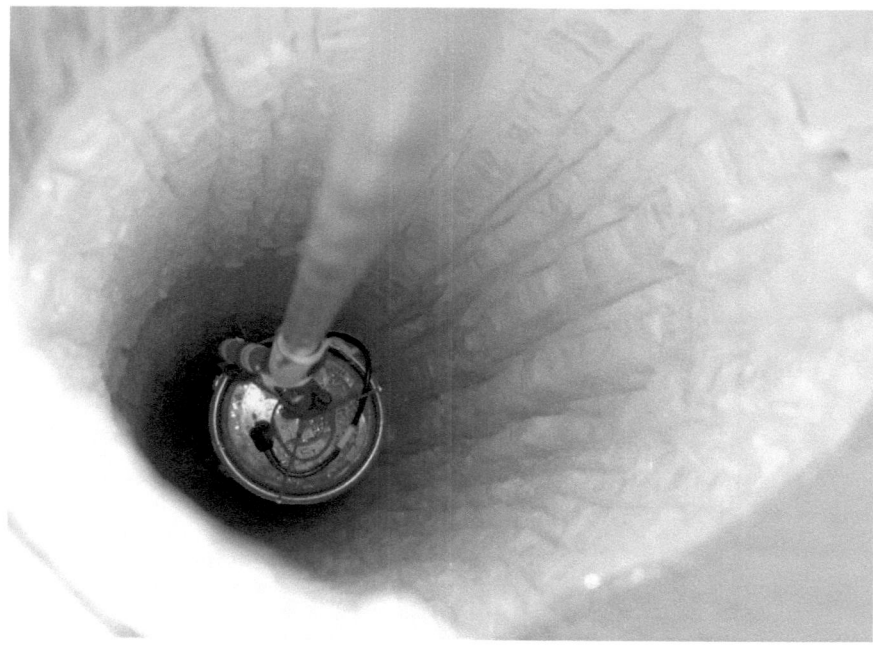

The second picture shows how an optical module is being sunk. The optical modules are the heart of the detector. The following image shows the internal structure of such a module:

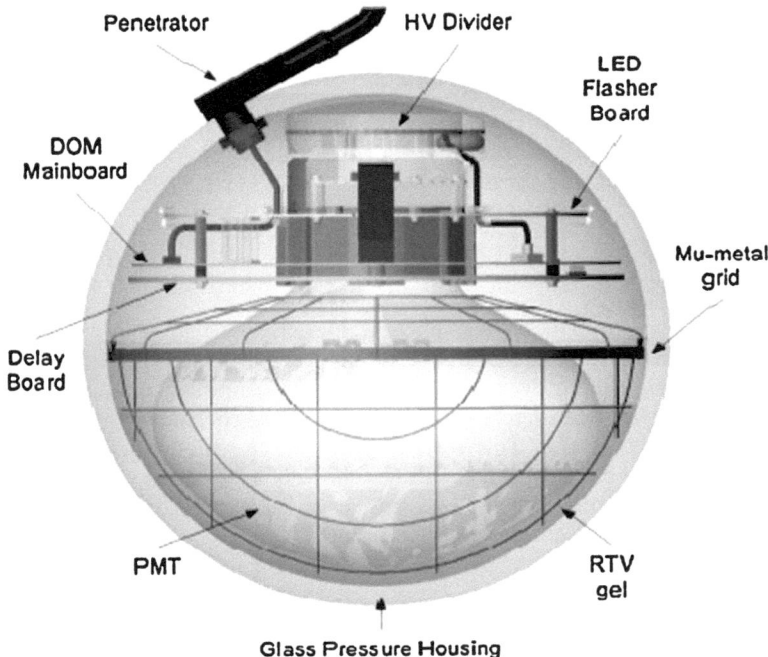

A brief and clear description of this structure is supplied e.g. by Jens Dreyer:

«A DOM [*digital optical module*] consists of a spherical glass case and the following components [...]:

- The *DOM-Main Board* is responsible for the digitization of the pulses and for the communication of the DOM. It has a clock that is synchronized with a global GPS Clock. Thus, each digital pulse can be signed with a time mark.

- A *Delay-Board* provides for a delay of the analog PMT [photomultiplier] signal. This is necessary so that the data acquisition can be started on time.

- In the *Flasher Board* there are a number of LEDs [light-emitting diodes], which serve for the calibration of the detector. In the Flasherboard there are also connections for additional sensors such as acoustic sensors. These can be integrated into IceCube, without changing the design of the DOM.

- Direct over the PMT is placed the *Baseboard*. In this component are accommodated the voltage divider and a toroidal transformer, which is required for the decoupling of PMT pulse and high voltage.

- The *High-Voltage Module* (HV-Module) consists of a control plate and a high voltage generator (HV-generator). In this component is generated the required high voltage for the operation of the PMTs, which can be adjusted individually from the surface for each DOM.

In the lower part of the DOM there is a Mu-metal screen that shields the DOM against the earth's magnetic field».[211]

It should be noted that the PMTs (photomultipliers) are placed in the lower part of the optical modules, and that they are oriented downward. The reason for this is that one tries to identify the neutrinos that have crossed through the whole earth with the IceCube detector, and which therefore are arriving «from the downside» in the detector. This way it is possible to remove a large part of the background noise caused by the particle showers of charged cosmic radiation right from the beginning, because in contrast to the neutrinos, these particles can penetrate only a few kilometers into the earth.

Another very important component of the IceCube detector is the ice itself, which determines substantially the properties of the observed events. The use of the Antarctic ice as detector medium for muon neutrinos using Cherenkov light has some advantages over other media (such as sea water).

Apart from the obvious stability of the experiment device, and the absence of organisms in the detector, which can trigger false events by bio luminescence, the relatively large absorption length of Cherenkov light in the ice represent one of the most important advantages of this medium. In the typical wavelength of Cherenkov radiation (300 to 400nm) the absorption length is between 50 and 150 meters. This implies that the Cherenkov light cone, caused by the movement of high-energy muons in the ice, will meet several DOMs when the muons pass through the detector.

However, the scattering length of light in the ice is about 20-40 meters, while, for example, the scattering length in the sea water is about 200-450. Since the scattering length in the ice is similar to the distance between the DOM in IceCube, it must be assumed that the received light is strongly scattered on his way to the DOM. This makes it more difficult to reconstruct the flight direction of the observed muons, and thus the angular resolution of the incoming neutrinos also becomes less accurate.

---

211 DREYER, J. (2005), *Hard- und Softwareentwicklung im Rahmen der Experimente AMANDA und IceCube* (Diplomarbeit, Technische Universität Dortmund 2005) 33-34.

Moreover, it turns out that the scattering length and absorption length within the ice volume vary widely. This is due to the air bubbles in the ice and the dust layers. The following illustrations show the absorption coefficient and scattering coefficient of the ice volume of IceCube as functions of wavelength and depth:

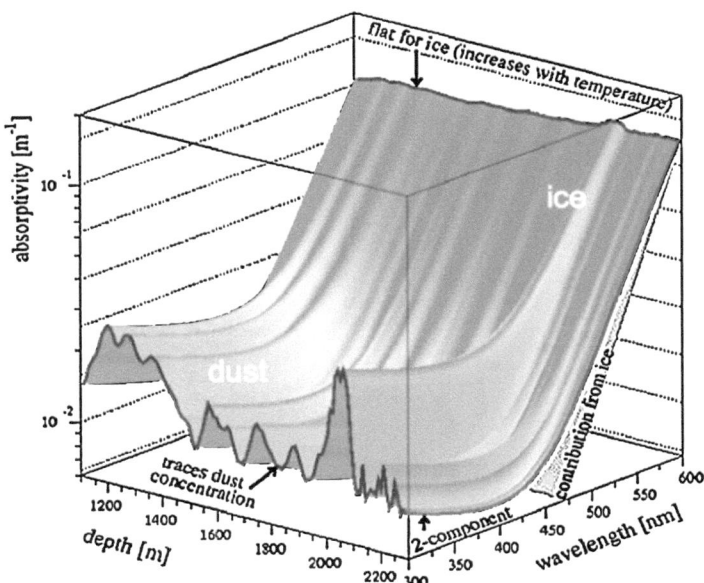

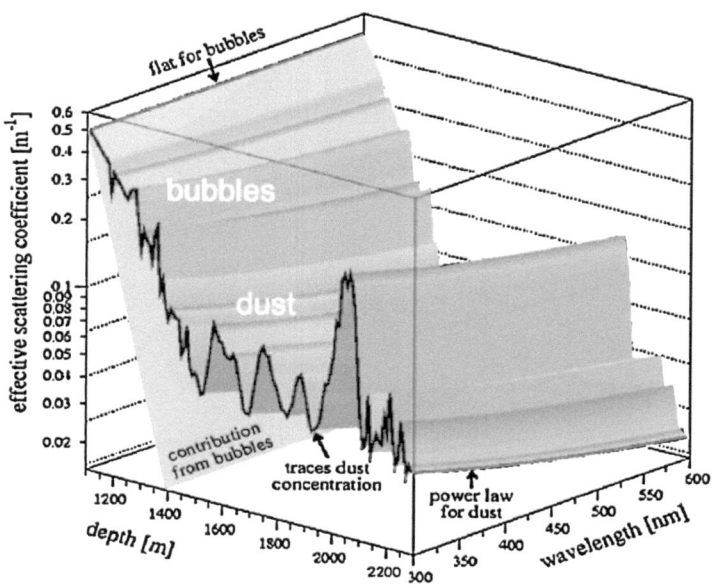

These illustrations show that although on average the two coefficients become always lower with increasing depth, the scattering and absorption properties of ice are very complicated in detail. (Especially the major layer of dust in a depth of 2000 meters causes significant deterioration of the values of the absorption and scattering coefficients). The characteristics of the ice must be necessarily considered in the data analysis.

(2) The Deep Core Area
The lower energy threshold for the detection of neutrinos in IceCube is about $E > 100 GeV$. To reduce this energy threshold, one would have to place the optical modules closer together. But this way one should expect a much higher background noise level generated by atmospheric events. In order to make the rate signal/background somewhat more favorable, the energy range below 100 GeV has been largely excluded.

As IceCube's predecessor, AMANDA could detect neutrinos up to an energy of 50 GeV (although it was 100 times smaller than the current detector), it was originally planned to integrate the AMANDA strings into the IceCube detector. But because of the high electricity consumption of AMANDA, this facility was switched off in 2009. Therefore, it was decided to sink 6 additional strings with 60 DOMs per string in a very clear area of the ice volume of IceCube. This area is called the *deep core*. One expects this extension of the device to reveal information about low-energy events up to 10GeV.

(3) The IceTop Detector

The IceTop is an air shower array of the type described in subsection 3.4.1 of the 2nd Chapter. When completed it will consist of 160 ice tanks, each with 2 DOMs. Two such tanks are placed over each string of the IceCube detector. The IceTop detector has two functions: on the one hand it serves for calibration purposes, on the other hand it makes it possible to put a veto on certain background-caused events:

> «The solid surface above IceCube allows the possibility of a surface air shower array that can be used for calibration (by providing a set of tagged muon bundles), for veto of certain bachgrounds generated by large air showers and for cosmic-ray studies. A square kilometer array of suitably design detectors on a 125m grid has full efficiency for air showers from 10^{15} eV to 10^{18} eV. [...]
>
> Most of the shower that trigger the detector will be near the threshold energy of approximately 200 TeV. Air showers with this energy typically contain 1 to 10 muons with sufficient energy to reach the deep detectors within a radius of 20-30 m of the shower trajectory. A fraction of these events will be used for calibration of the angular response and reconstruction algorithms of Ice Cube [...]
>
> In addition to providing a sample of events for calibration and to study of air-shower induced bachgrounds in IceCube, the surface arry will act as a partial veto. All events generated by showers with $E > 10^{15} eV$ can be vetoed when the shower passes through the surface array. In addition, higher energy events, which are a potential source of background for neutrino-induced cascades, can be vetoed even by showers passing a long distance outside the arr-ray. In particular showers with $E > 10^{17} eV$ and cores a km outside the array will produce a minimal trigger near the boundary of the array»[212].

The following image shows two ice tanks placed over a string[213]:

---

[212] «IceCube Preliminary Design Document». In:
http://www.icecube.wisc.edu/icecube/static/reports/IceCubeDesignDoc.pdf p.100-103.
[213] Source: http://gallery.icecube.wisc.edu/external/12-icetop/IMG_0641_JPG.jpg.html (reproduced with permission).

4.3 The Analysis of the IceCube Data

In the last chapter we have seen that the method of analysis of the MAGIC data is largely standardized. This also applies to other collaborations that work with the large gamma telescopes. However, it doesn't apply to the IceCube collaboration. The reason for this is that neutrino astronomy is in its early stages. Currently, various methods for the analysis of the IceCube data are being examined to see with which procedure one can achieve the most accurate results. In addition, it must be noted that IceCube is much more complex than the typical gamma telescopes with regard to three aspects: (1) the properties of the detector itself, (2) the functions of the detector, and (3) the rate signal/background in the detector.

Concerning the first point it should be noted that IceCube is a three-dimensional detector and not a two-dimensional one, as the gamma telescopes. One additional dimension implies that more parameters are to be considered in the reconstruction of the tracks and the energy of the observed particles, and that therefore the reconstruction algorithms are more complicated. In addition, it must be kept in mind that the detector medium (the ice) is not completely regular. The different concentration of air bubbles and dust in different places causes significant spacial variations regarding the scattering and absorption length of the detector. These aspects must be considered in the data analysis.

Regarding the second point it should be noted that one expects in principle to detect signals of very different origins and with very different characteristics in the IceCube. The search for neutrinos from GRBs must be organized in a different way than e.g. the search for neutrinos from the annihilation of neutralinos in the sun. And this, in turn, is to be accomplished in a different way than the search for evidence for neutrino oscillations or the search for traces of exotic particles. This means that different strategies for the data analysis must be examined in the different research projects.

Whith regard to the third point it must be noted that the rate signal/background in IceCube (or more generally in the neutrino detectors) are much worse than the corresponding rate in the gamma telescopes. We have mentioned in the last chapter that, for instance, the rate signal/background in the case of MAGIC is about 1/1000. In the case of IceCube, however, the rate lies somewhere between $1/10^6$ and $1/10^8$. That means that from one million events detected in this device only one is relevant. This represents a major challenge regarding the procedures for the separation of signal and background. And consequently it is not surprising that new separation algorithms are tested again and again.

But despite these circumstances it is important for the purposes of our study to outline, the typical data analysis at EisCube. For one can only say to what extent the data analysis of IceCube leaves room for constructivist developments with the help of this information.

Let us regard a typical case: the search for muon neutrinos from GRBs with IceCube, as it is e.g. sketched in Abbasi et al. (2009a), (2009b) and other papers. (The search for neutrinos from AGNs or from other cosmic point sources runs very similarly).

One starts the search with a large amount of raw data, which were recorded as potential events. In order to classify the signals from the DOMs as events some prior requirements must be fulfilled. It is specifically required that a minimum of 8 neighbouring DOMs within a time window of 5 µs send a minimum signal of 0.25 photoelectrons. One can safely assume that the signals that do not satisfy these trigger requirements are not caused by the muon neutrinos searched for.

However, most triggered events are caused by muons originated from particle showers caused by charged cosmic radiation in the Antarctic atmosphere over the detector. At this point the rate signal/background is (as just mentioned), somewhere between $1/10^6$ and $1/10^8$. To minimize this tremendous background the earth can be used as a background filter. The reason of this is that most of all neutrinos (except for extremely high-energy neutrinos) can pass through the earth without any interactions, while the cosmic radiation only penetrates the earth to a depth of a few kilometers. One can therefore achieve a strong background cleanup of the data by selecting only those events that are identified in a first track-reconstruction as particles moving bottom-up in the detector.

In order to find these events, one uses different algorithms for track reconstruction. Often just a line-fit algorithm is used, which provides a fast track reconstruction based on the measured hit times in the various DOMs[214]. This is a simple and not very accurate method (because e.g. the scattering effects of the Cherenkov light in the ice are not considered) but it is good enough to drastically reduce the initial data. In the data already recorded with the first 22 strings of the IceCube usually only a few thousands neutrino candidates remain from the raw data gathered in hundred days.

However, even in this reduced data the background exceeds the looked-for signal by several orders of magnitude. The remaining background consists of three components: (1) particles with incorrectly reconstructed track: some particles that move top-down through the detector, might also be classified as particles that move bottom-up by the line-fit algorithm, (2) coincidence events: two atmospheric muons that cross top-down in rapid succession the same sector of the detector, resemble a bottom-up event, and (3) atmospheric neutrinos: neutrinos, produced by particle showers on the other side of the earth cross the detector bottom up, and if they produce muons, the trace of these muons is the same as the trace of the neutrinos produced by cosmic muons.

To extract the components (1) and (2) of the residual background, a series of quality-cuts is applied to the events . The usual parameters defining these cuts can be defined are the following:

«– θrec: reconstructed zenith angle;

– σdir: the uncertainty on the reconstructed track direction (quadratic average of the minor and major axis of the 1σ error ellipse);

– Lred: log of the likelihood value of the reconstructed track divided by the number of degrees of freedom (number of hit DOMs minus number of fit parameters). This has proven to be a powerful variable for separating signal and background as visible from Figure 2;

– LU/D: difference in log-likelihood value between the reconstructed track and one containing a bias to be reconstructed as downgoing. The bias is zenith angle dependent and follows the rate of downgoing atmospheric muons. The rationale behind this is that a track is much more likely to originate from an atmospheric muon than from a muon generated in a neutrino interaction. Only high-quality upgoing tracks have high LU/D values;

---

[214] An explanation of this and other reconstruction algorithms can be found in AHRENS et al (2004). For more information about these algorithms see e.g. WIEDEMANN (2008) chap.5.

– Ndir: the number of photons detected within a -15 to +75 ns time window with respect to the expected arrival time for unscattered photons from the muon track hypothesis;

– θmin: minimum zenith angle from a fit of a two-track hypothesis to the light pattern. For this, the light pattern is divided into two separate sets of hits based on the mean hit time of the event, and a track hypothesis is fitted to each hit set separately. A cut on the smaller zenith angle, θmin, of the two tracks is very effective against so-called coincident muons, where two muons from different atmospheric showers pass the detector in fast succession mimicking an upgoing track. In this case, often at least one of the two tracks is reconstructed as downgoing whereas for good-quality upgoing neutrinos both tracks appear most often as upgoing; and

– Erec: reconstructed muon energy at point of closest approach to the center of gravity of hits in an event (Zornoza et al. 2008)»[215]

To select the best values for the cuts in these parameters, computer simulations are used in a similar way as already discussed for the data analysis of MAGIC. The main steps of the simulation chain are the following:

«–At the beginning of the simulation particles must be generated. This is done by the generators and CORSIKA NuGen.

–The propagation of the resulting charged leptons through the earth and the ice and the energy losses of the leptons are the next simulation steps. The spread of the Cherenkov light and the resulting number of Cherenkov photons in a DOM can be determined. However, this is not done via simulation, but by previously calculated tables the «photonic tables», or short «photonics».

–In the PMT simulation one simulates how many photoelectrons are generated by the incident Cherenkov light and how the shape of the resulting wave looks like.

–The DOM simulation simulates the behavior of the other DOM components, especially the ATWD and FADC.

–Finally, the triggers are simulated, which are needed for each analysis».[216]

After testing some cuts in the space of the mentioned parameters, Abbasi et al. applied the following cuts in their data:

---

[215] ABASI et al. (2009b) 12-13.
[216] CLEVERMANN (2009) 29-30.

$\theta_{rec} > 85°$; $\sigma_{dir} < 3°$; $\theta_{min} > 70°$; $L_{U/D} > 30$; und

$$L_{red} \leq \begin{cases} 7.8 \text{ for } N_{dr} < 7 \\ 8.5 \text{ for } N_{dr} = 7 \\ 9.5 \text{ for } N_{dr} > 7 \end{cases}$$

It must be noted, however, that these numbers must again be determined individually for each record. Because which cuts obtain the best background suppression depends on the specific conditions under which the data was gathered. At this point it can also be interesting to mention that just as there is a variant of the data analysis of MAGIC, in which the random forest method (instead the cuts) is used for the separation of signal and background, the IceCube Collaboration currently also tests whether with the help of the random forest method a better background suppression can be obtained. However, since the development of this variant of the analysis has not yet been completed, it is ignored here.

After the cuts practically only events caused by neutrinos, which reach the detector from the northern hemisphere of the earth remain within the data. In other words, the incorrectly reconstructed events and coincidence events have been extracted. However, most of the neutrinos that reach the detector from the northern hemisphere are not cosmic neutrinos but atmospheric neutrinos produced in the particle showers caused by the charged cosmic radiation in the atmosphere of the northern hemisphere. This last source of background can not be extracted with additional cuts, since there basically is no difference between the events caused by cosmic or by atmospheric neutrinos.

What can be done here? Erik Strahler summarizes the just described analysis strategy and the steps that still can be taken to extract the signals from the background of atmospheric neutrinos as follows:

> «Searches for astrophysical neutrinos face significant backgrounds. Muons created by cosmic ray interactions in the atmosphere dwarf the signal by six to eight orders of magnitude, even at a depth of several kilometers. From the opposite hemisphere, however, this muon background vanishes. We therefore utilize the Earth as a filter to form a sample of muons induced by neutrinos.

The reconstruction process is not perfect though, and we must apply a series of quality criteria to remove the misreconstructed donwgoing muons from our data sample. Even after this selection, we are left with an irreducible background of muons from cosmic ray-induced neutrinos. In order to distinguish these from the astrophysical signal we must either search for clustering in the sky, associations with known point sources, or take advantage of the differences in energy spectrum. Atmospheric neutrinos follow an $E^{-3.7}$ power law, while the signal spectrum is typically harder, depending on the details of the model.»[217]

Now it is about either showing that in certain directions there is a significant surplus of neutrinos over the background of atmospheric neutrinos or that the entire energy spectrum of the investigated neutrino sample is flatter than the spectrum of atmospheric neutrinos.

To find possible neutrino surpluses, one currently uses the so-called «binned analysis» and «unbinned analysis». Elisa Bernardini has summarized this approach as follows:

«*The IC22* [IceCube 22strings] *binned analysis* distinguishes a localized excess of signal from a uniform bachground using circular angular search bins. The sensitivity (average upper limit in case of no signal), in a given declination band, can be obtained from Poisson statistics and it is calculated by comparing the number of simulated signal events for a given flux Φ^0 reconstructed inside the search bin, n_{sig}, to the average number of background events as measured from scrambled data, n_{bg} [...]

The IC22 un-binned analysis construct a likelihood function which depends on the signal probability density function (pdf), obtained from MC simulation of the event energy and angular distribution, the background pdf (obtained from data as a function of the event position x_i), and total number of data events n_{tot}»[218]

For the search of a flattening of the power spectrum of the examined neutrino collection one uses a different method. First one selects a series of parameters, with the help of which the energy of the neutrinos of the collection can be reconstructed. Some of these parameters can be combined in a neural network. The

---

[217] STRAHLER (2009) 2.
[218] BERNARDINI (2009) 5.

combined parameter and the others are then used as the basis for an unfolding algorithm[219] the output of which is the energy of the neutrinos[220].

Finally, it should just be noted that none of the variants of the data analysis has provided evidence for the existence of cosmic neutrino point-like sources so far. That is, at present all the results are consistent with the hypothesis that the neutrinos observed by IceCube are all atmospheric neutrinos. But one still expects that (hypothetical!) cosmic sources finally become visible when the whole detector is complete and in operation.

4.4 IceCube and AMANDA

In order to complete the presentation of the work of the IceCube Collaboration it is appropriate to draw attention to the prehistory of this experiment. For both, the presented instruments and the just described method for data analysis, are the first results of a lengthy development starting with the design of the AMANDA Detector at the beginning of the 1990s.

In 2004, the AMANDA Collaboration consisted of more than 100 scientists, from nineteen research institutions in America and Europe[221]. This collaboration pursued two goals:
(1) The first one was to acquire the necessary technical expertise construct a neutrino telescope in the Antarctic ice with a volume of $1 km^3$ at a later stage,
(2) the next one was about testing different theoretical models according to which some sources of cosmic neutrinos should be detectable with the help of a detector with an effective area of $10^4 m^2$. The AMANDA Detector would reach, and even exceed such a surface[222].

The AMANDA Detector was built at the South Pole (close to the current location of the IceCube) between 1993 and 2000. The idea of a system of strings with optical modules, which is to be sunk into the ice was first realized by AMANDA.

First, four strings with 20 optical modules per string were placed at a depth between 800 and 1000m ["AMANDA-A"] in the season 1993/1994. However, it was discovered that there are so many air bubbles in this depth that the

---

[219] Some details about such algorithms can be found in BLOBEL (1985), and BLOBEL (1996). An updated version of this unfolding program is being currently developed by Natalie Milke at the Technische Universität Dortmund.

[220] A detailed description of this procedure can be found in MÜNICH (2007) chap. 5-6.

[221] See: http://www.amanda.uci.edu/collaboration.html

[222] The effective area of the AMANDA II detector is $3 \cdot 10^4 m^2$.

dispersion of the Cherenkov light caused by the bubbles prevents the track reconstruction of the incoming particles.

After a thorough investigation of the properties of the ice in the vicinity of the device the decision was made to place a detector at a much deeper level. In 1995/1996 four strings (again with 20 optical modules per string) were placed at a depth between 1545 and 1978m. The new detector was called AMANDA-B4. Placing the system at a much deeper level significantly reduces the problem with the bubbles. However, it was realized soon that smaller distances between the modules would simplify the data analysis.

Then the placement of six additional strings followed [AMANDA-B10] in 1997/1998. In this season the number of modules was increased to 36 per string, in order to reduce the distances between the modules. But again the need for further changes in the design of the device was identified. First of all, the manufacturer of optical modules had to be changed. For the previously manufactured glass beads were partially polluted by radioactive ^{40}K which increases the noise in the detector. But the problem with the data transfer was even more serious: up to this point the data of the optical modules was transmitted to the surface by analog electrical pulses. However, this method had the disadvantage that analog electrical signals dispersed very strongly during transmission. And so it happened, for example, that electrical pulses with a length of 20 ns within the modules reached the surface with a length of 200ns.

To avoid this serious data loss, the design of the optical modules was changed: the next strings were provided with modules which can lead the data to the surface electrically as well as optically. Between 1997 and 2000 the development of the detector was finally completed. In its final configuration, AMANDA –now called AMANDA-II– consists of 19 strings, with a total of 677 optical modules. In the eighteenth string another modification of the mode of transmission of the signal was tested. The modules of this string were designed in such a way that the received signal is digitized first and then transmitted. After the examination of the operation of the digital modules it was decided that this would be the variant to be used for IceCube.

The following figure summarises the construction of the AMANDA detector, up to the AMANDA-10 stage[223]:

---

[223] Source: http://www.amanda.uci.edu/images/AmII_eiffel.gif (reproduced with permission).

192 4. Chapter: IceCube

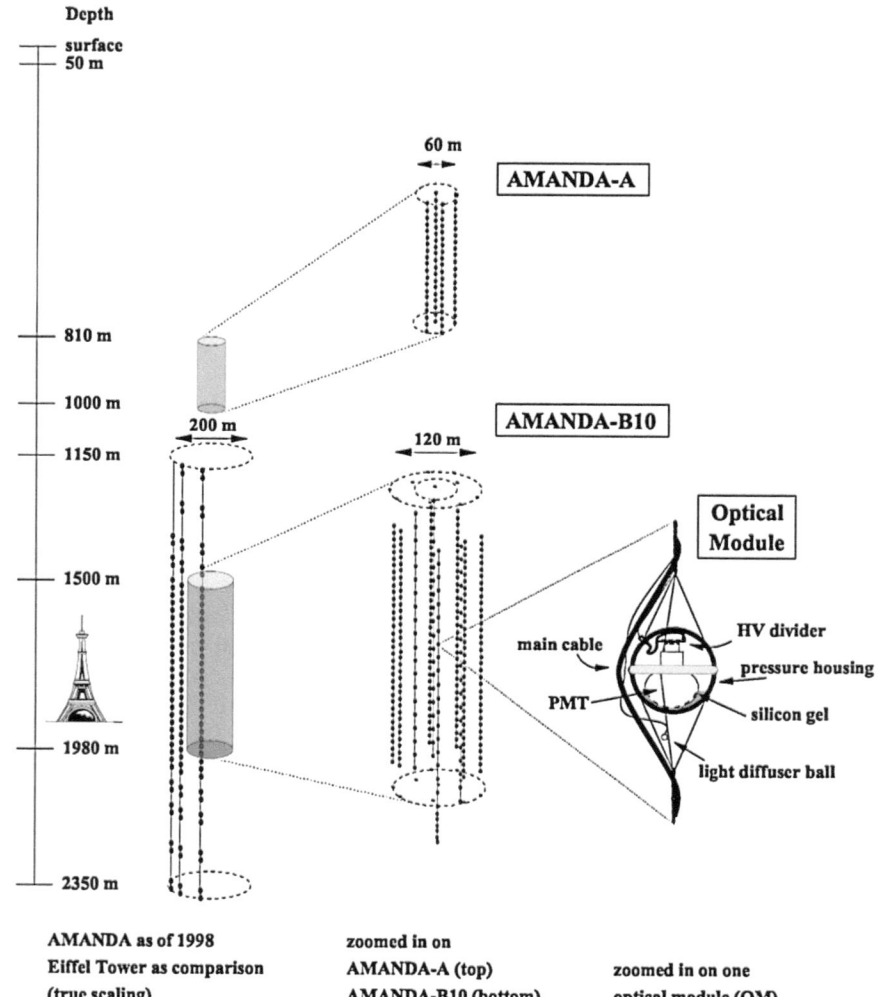

AMANDA as of 1998
Eiffel Tower as comparison
(true scaling)

zoomed in on
AMANDA-A (top)
AMANDA-B10 (bottom)

zoomed in on one
optical module (OM)

AMANDA provided data until 11 May 2009. Although the detector was still functional, it was shut down to save energy. During its eight years of operation the data from AMANDA led to strict upper limits for possible neutrino flux from different cosmic sources. No significant source has been observed up to now.

5. IceCube and the Debate between Realism and Social Constructivism

Let us once again consider the questions presented as the key questions at the beginning of the chapter:

What happens when the last of control level for the work of scientific collaborations – the examination of the results by other collaborations with similar instruments – is missing? Will this case factors such as, for example, the pressure to succeed (to justify the project costs) or the aspiration of scientists to make a career and earn fame, inevitably lead to dynamics of research in the sense of social constructivism?

The research of the IceCube collaboration gives us an example of the work of collaborations owning a unique detector. Because there currently is no other neutrino detector of this size and type, and similar research facilities in Antarctica are not planned at present. It is possible that in some years the current initial attempts to build a large neutrino telescope in the Mediterranean will succeed. But the success of these experiments is by no means guaranteed, and such a detector would have very different properties than IceCube – if just for the reason alone that the properties of the Antarctic ice and the sea water as a detector medium are very different–.

Does that mean that future results and successes of the IceCube experiment are pre-programmed from the beginning? Or is it more likely that the research of the IceCube collaboration is unbiased as to the result?

After the presentation of the last section one can certainly get the impression that the methods of the IceCube collaboration can basically be regarded as reasonable in the sense of a realistic and open-ended search for new physical phenomena. Some of the reasons are that the principles of the detection methods [Cherenkov light] and the corresponding devices [optical modules with photomultipliers] are well known and that the practical implementation of these principles in the detector is the result of many years of development[224] due to numerous calibration tests[225] for the individual components of the detector as

---

[224] Interesting examples of such development can be found e.g. in DREYER (2005).

[225] To avoid that the issue of the previous chapter is repeated again, I have not reported in detail about the arduous process of calibration of the DOMs. It should only be mentioned here that the calibration of each new string runs over a period of more than one year in the various tests are performed, some of which are similar to the calibration tests for the pixels of the MAGIC camera.

well as for the whole detector and that a number of standard procedures for the data analysis (such as the «quality cuts», the «Monte Carlo Simulations», the «Random Forest» Method, etc.) are applied, leading to good and reliable results in other areas of physics.

However, at the beginning of this chapter it was pointed out that –according to the constructivist view– one should not pay so much attention to such experimental and rational reasons, even if they are cited in scientific publications to support the results obtained. That is, one should (from a constructivist point of view) not overestimate the importance of the calibration tests, the «cross checks», the comparisons between data and simulations, etc., as determinants for the dynamics of research. Because, strictly speaking, each explanation for the development of a certain research line by referring to such elements only represents an *a posteriori* rationalization strategy, while the crucial reasons for the development of the research are extra-scientific social factors like e.g. career chances, the chances of receiving state funds, the intellectual trends, etc.

Such an opinion can therefore not be rejected just by referring to the technical justifications of research results, as the constructivists question the importance of these technical justifications from the outset. In other words, it certainly was appropriate to present the experimental and rational reasons that physicists regard as the determining factors for their results in the previous section (and even in more detail in the previous chapter). But this presentation alone is not enough to eventually invalidate the constructivist logic. After a constructivist had taken notice of the entire range of calibration procedures, checks and analysis procedures outlined in the last chapter he can, for example, argue that all these controls can be multiplied in principle *ad infinitum*: one may consider it necessary to examine the checks, and the checks of the checks etc. And so, the final decision about when the results are really secured must be extra-scientifically motivated.

In my opinion, such a position is unconvincing. The chain of checks certainly has to stop at some point so that results can be presented. But the checks can go so far that there no longer are reasonable doubts regarding the results. However, this is an argument based solely on common sense. Consequently, it is no fundamental argument. In principle, the statement that the chain of checks of each investigation must be stopped at some point by a decision is correct. Therefore, it can basically always be questioned whether the point where such a decision is made was possibly determined by extra-scientific interests.

Well, if the reference to the technical justifications of research results is not enough to decide the controversy between realism and social constructivism concerning the dynamics of research, what can we do to break this impasse?

I propose that we reflect upon the future development of the IceCube project, assuming that this development follows constructivist dynamics. I also propose to consider the development of the AMANDA Project from the same perspective. The first task is easy. On the one hand we need to remember that the IceCube Project has cost more than $ 300 million so far. And on the other hand we must look again at the list of motivation points for the construction of the detector, already quoted at the beginning of the previous section:

«A well-designed neutrino telescope can

– search for high energy neutrinos from transient sources like Gamma Ray Bursts (GRB) or Supernova bursts;

– search for steady and variable sources of high energy neutrinos, e.g. Active Galactic Nuclei (AGN) or Supernova Remnants (SNR);

– search for the source(s) of the cosmic-rays;

– search for Weakly Interacting Massive Particles (WIMPs) which may constitute dark matter;

– search for neutrinos from the decay of superheavy particles related to topological defects;

– search for magnetic monopoles and other exotic particles like strange quark matter;

– monitor our Galaxy for MeV neutrinos from supernova explosions and operate within the worldwide SNEWS triangulation network;

– search for unexpeted phenomena.»[226]

Furthermore, we can also consider the concluding remarks of the chapter «Science Motivation for Kilometre-Scale Detectors» from the «IceCube Preliminary Design Document»:

«Discovery of any single one of the high energy signals listed above would unquestionably make IceCube a resounding success. [....] [As] a detector one hundred times larger than AMANDA and one thousand times larger than any underground detector, IceCube will be opening a new window on the universe»[227]

When considering both, the material cost –to be justified– and the big advantages for the careers of all scientists involved in the project, it should be expected from a constructivist point of view that at least some of the expected phenomena will be discovered –or rather constructed!– in the next years. But this also means that without the expected phenomena (or other completely new phenomena) being

---

[226] «IceCube Preliminary Design Document». In: http://www.icecube.wisc.edu/icecube/static/reports/IceCubeDesignDoc.pdf p.12.
[227] Ibidem 25.

discovered at the end this result can hardly be understood from the constructivist perspective[228].

It is still too early to say whether the story of the IceCube collaboration will be a prime example for the failure of social constructivism to explain the actual dynamics of research in physics. But the story of the AMANDA experiment could fairly be regarded as such an example, even if not so outstanding. Let us look at this example more closely:

In the last section it was pointed out that the AMANDA collaboration consisted of more than 100 scientists from nineteen research institutions. It also was a large scientific collaboration funded with several millions dollars. It was also pointed out that the AMANDA collaboration pursued two objectives:

(1) to win the necessary technical experience, in order to construct a neutrino telescope in the scale of $1 km^3$ at a later stage; (2) to test various theoretical models, according to which some sources of cosmic neutrinos should be detectable using a detector that has an effective area of $10^4 m^2$.

In 1996 –when the first four strings of the AMANDA-B detector were placed at a depth between 1545 and 1978m– we find e.g. the following reference to these goals in an article by Francis Halzen (for the AMANDA collaboration):

«Speculations that the highest energy photons and protons are produced by cosmic accelerators powered by the supermassive black holes at the center of active galaxies can be used to estimate the required effective volume of a neutrino telescope. The answer is $1 km^3$. [...] Model building suggests that the detection of these accelerators may be within reach of much smaller detectors with effective area of order $10^4 m^2$. The AMANDA collaboration is ready to complete such an instrument within the next few months»[229]

And in the same paper paper Halzen also points out that even the indirect detection of WIMPs («Weakly Interacting Massive Particles») is to be expected. Be-

---

[228] Unfortunately, from the discovery or no-discovery of the previously mentioned phenomena, no conclusion can be drawn regarding the realistic perspective, because the final outcome of the research with the IceCube (considered from a realistic viewpoint) is open. But I have for example in the 2nd chapter of this study also mentioned that other possible developments in the field of astroparticle physics might cause problems for the realism: If, for example, in the near future, a broad consensus is achieved that the data of the observation of the gamma flares of the AGNs demonstrate the violation of the Lorentz invariance, without first understanding in detail the exact sources of these emission processes, and other aspects, which could lead to alternative explanations of such observations, we would have found a development that has to be interpreted in a constructivist rather than a realistic manner.

[229] HALZEN (1996) 1.

cause in most theoretical scenarios the neutrino flux associated with the annihilation or decay of WIMPs should lie within the observable range of the AMANDA detector.

Such optimistic expectations can be found in all papers of the AMANDA collaboration from that period:

«Who knows what secrets will be uncovered as the wealth of data from AMANDA-II is processed and analyzed?»[230]

This sentence can still be found on the web page of AMANDA-II. And although one of the goals of such formulations is to draw public attention to one's own work, it can easily be imagined what high expectations the participating scientists had for the experiment. However, in the meantime everything points to the fact that the expected great discoveries the AMANDA detector was hoped to make are missing. Instead, the results achieved after a decade of data acquisition and data analysis can be summarized as follows:

«Results from AMANDA-II [...]:

1. A search using AMANDA-II established a flux limit for point sources.

2. There is no evidence for any neutrino point sources, although some interesting, but low-statistics, possibilities bear further investigation.

3. AMANDA-II has provided an extremely useful test module for the much larger IceCube [...]»[231]

In other words, none of the expected neutrino and other signal sources were detected, no AGNs, GRBs, WIMPs or exotic particles. After a thorough data analysis the members of the AMANDA collaboration were only able to set upper limits for the neutrino flux from these sources. And these limits imply that the neutrino flux should be much lower than predicted by established theoretical models in the 90th. Here is one example:

At the time when the AMANDA detector was built, several authors proposed models, according to which the emission of x-rays from the jets of radio-weak AGNs is correlated with the emission of neutrinos. As the x-ray satellite ROSAT has been in operation since 1990 and provided the necessary data about the x-ray radiation from many AGNs, the expected neutrino flux from this kind of source could be calculated by means of theoretical models. As a result of these calculations it was predicted that the AGN neutrinos should be detectable by flattening the neutrino spectrum in the high-energy range; a phenomenon which could be already discovered with AMANDA.

---

[230] Source: http://www.amanda.uci.edu/public_info.html
[231] BOYD (2008) 82.

At this point it should not be forgotten that the high-energy range is just the most difficult to examine (due to the inverse correlation between the increase in energy and the neutrino flux). That is, if the members of the AMANDA collaboration were willing to construct a suitable result, they would have had a relatively easy job with the data analysis.

However, the predicted flattening of the neutrino spectrum has not yet been discovered, and many theoretical models have therefore been rejected. The following figure, for example, shows the comparison between the predictions of four theoretical models and the analysis of the data collected by AMANDA between 2000 and 2004[232]:

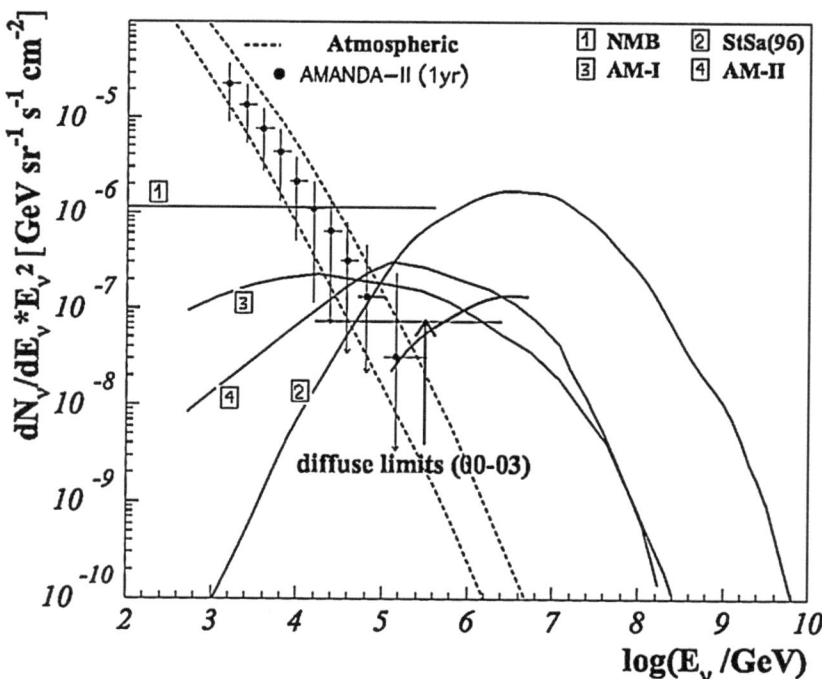

---

[232] Credit: BECKER et al. (2007) .26. The theoretical models whose predictions are compared with the data in the picture are the following:
[1] NELLEN et al. (1993).
[2] STECKER – SALAMON (1996).
[3] und [4] ALVAREZ-MUÑIZ – MÉSZÁROS (2004).

Two pieces of information can be taken from this figure: first of all, that the represented theoretical models are inconsistent with the upper limits of the neutrino flux that can be derived from the AMANDA data, and secondly, that so far the AMANDA data are consistent with the hypothesis that all observed neutrinos are atmospheric neutrinos –or in other words that there is no significant flux of high-energy cosmic neutrinos–.

The search for neutrinos from GRBs or from the decay (and/or annihilation) of WIMPs has been leading to the same negative result so far.

Every time I have discussed this result with experts, their opinion was that this is a normal development. Either the expected phenomena are actually there or they are not. And the experiments are being performed in order to test whether the physical reality behaves the way we described it in theoretical models. However, one must necessarily ask how such development can be interpreted from a constructivist perspective. For we have mentioned the well-known verdict of Pickering concerning the discovery of neutral weak currents in the first chapter of this study:

> «Quite simply, particle physicists accepted the existence of the neutral current because they could ply their trade more profitably in a world in which the neutral current was real»[233]

Well, if dynamics of research are largely driven by profit, how can the negative results of AMANDA be explained? For all the profits in this case are clearly on the side of confirming the theoretical expectations. And what is at stake here is nothing less than the opening of a new observation window for the investigation of the universe.

Possibly, it could be argued at this point that social constructivism does not claim each theoretical idea can be proved[234]: in the process of constructing phenomena it is possible to find «resistances» on the part of the material world at times, which must be met by «adaptation» on the part of the physicists.

However, a central thesis of the constructivist view on dynamics of research is that when a theoretical scenario is attractive enough for both, theorists and experimentalists, a way to overcome the obstacles of nature will eventually be found, and experimental evidence for the desired scenario be delivered. Because otherwise nature would be the crucial instance in the process of constructing the phenomena. And such constructions could rather be seen as representations of nature in the sense of realism. Due to this reason it can be claimed that the results

---

[233] PICKERING (1984b) 87.

[234] Although there are constructivists such as Barnes, who say just this:
«Reality will tolerate alternative descriptions without protest. We may say what we will of it, and it will not disagree. Sociologists of knowledge rightly reject realist epistemologies that empower reality» BARNES (1991) 331.

of the AMANDA collaboration can hardly be understood from the constructivist perspective.

On the other hand is the development of the research with the IceCube detector still open: can the existence of the expected cosmic neutrinos finally be proved with the help of the new detector? Or will the data from IceCube also be consistent with the hypothesis that only atmospheric neutrinos can be observed with this instrument?

From a realistic point of view, the scientists who currently work on the development of IceCube see this question as completely open. But it is obvious that only the confirmation of theoretical expectations is consistent with the constructivist approach. If at the end this confirmation is missing (as in the case of AMANDA) there will be a good example for social constructivism not being able to provide an adequate explanation for dynamics of research. In this case it would have been proven that social factors –such as the pressure to justify the financial support of a collaboration or other social factors as the incentives associated with finding empirical evidence for a promising hypothesis before other research teams do it, etc.– are not suitable to determine dynamics in a typical field of research (as outlined here).

In a previous section, we have mentioned the analysis of López Corredoira of the work of the current large scientific collaborations:

«[...] science is bought, and the one who pays has the right to demand the fruits are superior to those obtained for a lower price. [...]

Spontaneity has no place in actual science. Neither, is there a place for fortuitous discoveries favouring those who suspect that something in astrophysics is not following an appropriate path. Almost everything is planned in order according to programming and forecasts made many years in advance»

However, the experience of the AMANDA Collaboration argues against this evaluation and it is possible that the future work of the IceCube Collaboration gives reason for new objections. The consequence would be that in the current research of this collaboration not everything is purchasable and constructible and that planning results in advance is no guarantee for these results being actually obtained in experiments. The development of experimental physics –including the work of large collaborations– cannot be planned on paper because nature always has the last word.

5. Chapter: Conclusions and Outlook

1. Introduction

In the previous three chapters we discussed many details of the experiments in astroparticle physics. In particular, different aspects were focused on in connection with the MAGIC and IceCube collaborations –their goals, work methods, analysis procedures, applied techniques–. To complete our study we now summarize the impulses for the discussion between realism and social constructivism, gained from the research in the field of astroparticle physics.

These impulses have to a large extent already been suggested in different passages of this book. Therefore, the primary issue here is to classify them in order to get an overall view of the philosophical content of this study. Since no further analysis of the experiments must be accomplished, this chapter will be much shorter than the previous ones. However, a simple summary of the relevant points from the previous pages is not sufficient in order to extract the philosophical conclusions from the collected materials. Possible objections regarding the conclusions drawn (or at least suggested) on the basis of these materials in the course of our study must be discussed. In addition, some considerations regarding the scope of the results are also necessary.

To this end, the chapter is structured as follows: first, [2nd Section] a preliminary answer is given on the basis of the examples presented in this study concerning the question of the suitability of the constructivist and realistic perspectives to describe the dynamics of the research within the field of astroparticle physics. Then [3rd Section] some examples of possible future developments in research are summarized, which could confirm (or question) this preliminary answer. Subsequently [4th Section], the conclusions drawn here are examined critically. Finally [5th Section], the question is discussed to what extent the results in the field of astroparticle are transferable to other fields of physics, and [6th Section] the consequences this transfer has for the discussion about the rational order of nature are elucidated.

2. Remarks about the Social Constructivist and Realistic Interpretations of Dynamics of Research within the Field of Astroparticle Physics

Based on the general description of the experiments in the field of astrophysics in the 2nd chapter, and especially on the much more detailed presentation of the work of the MAGIC collaboration and the IceCube collaboration in the 3rd and 4th chapter, one can draw the conclusion that dynamics of research in astroparticle physics is in a natural way realistic interpretable, while many details of these dynamics hardly appear understandable from a constructivist perspective. Two different analyses converge to this result: the analysis of the difficulties, which would have to be overcome with regard to a possible construction of the phenomena, and the analysis of the actual response of experimental physicists to «profitable» scenarios, which could be the result of the discovery or the construction of certain phenomena in the area of astroparticle physics.

The technical difficulties for the construction of the phenomena in astrophysics were mainly examined in the2nd and 3rd chapter. They can be summarized as follows:

In order to construct certain desirable phenomena in this area, one must arrange either the data taking or the data analysis in such a way that the expected results are the end product of the measurement method. However, none of these ways seem to be really open. Because in order to influence the record of the raw data, one would need some freedom to decide when the individual components of our measuring devices (or the whole devices) work correctly. But the detectors in astroparticle physics consist of calorimeters, spectrometers, photomultipliers, mirror systems, etc., used in different variations in many other experiments in physics and other sciences. That is, there is a lot of previous know-how about the performance, which will provide such equipments. Thus, the correct functioning of the detectors and their components is secured by standard controls, such as cross checks, calibration and monitoring of known phenomena. When dealing with the details of these controls there hardly is any space for influencing the raw data in such a way to obtain the desired end results.

Similar difficulties can also be found regarding the data analysis. One could assume *prima facie* that the data analysis can be influenced mainly in two places, namely when selecting the cuts used to eliminate the background, and when selecting the theoretical models of the processes the Monte Carlo Simulations (always used in data analysis) are based on. But if one examines these places more closely it doesn't become clear what kind of free control of the results might take place here. Because the cuts are not selected arbitrarily. Instead, one rather uses those cuts, which led to the best results in the analysis of data resulting from the Monte Carlo Simulations.

And with regard to the simulations most of them are unproblematic because they describe very simple (and/or well definite) sub-processes. Other simulations, however, are complicated and contain some uncertainties (such as extrapolations of cross sections, etc.). In the field of the ground experiments of astroparticle physics this particularly refers to simulations of the generation and spreading of atmospheric particle showers induced by cosmic particles. However, these simulations (in particular the various versions of CORSIKA) are developed by teams not bound to specific experiments in astroparticle physics. That is, the researchers of the collaborations that might be interested in constructing certain phenomena have the most complicated (and therefore the most appropriate for interference) simulation of the analysis chain already specified. Thus, this simulation cannot be designed to confirm any hypothesis.

In addition, a construction of certain phenomena, wittingly or unwittingly, is above all prevented by the fact that there usually are different collaborations, which are able to examine the announced new phenomena independently (and with different kinds of detectors). Under these conditions a constructed phenomenon could become generally accepted only when members of several independent collaborations are so fascinated by the theoretical prospects opened by this phenomenon, that all (or most) teams would interpret their own data in a similar way. But the fact is that new results announced as discoveries by any collaboration are always meticulously examined by other teams, trying to find all possible errors.

What we can learn from this is that the occurrence of constructivist dynamics of research is only to be expected when a large collaboration equipped with a new, unique detector carries out investigations which cannot be checked by other collaborations now or in the near future. If we, however, take into account at this point the analysis of the actual answer of the experimental physicists to the scenarios, in which the circumstances apparently favour the construction of certain phenomena, we find no indication for such a construction. Even under the most favourable circumstances the example of the AMANDA collaboration mentioned in chapter 4 show that dynamics of the research are rather different to what constructivist would have expected. That is, the study of the actual reaction of experimental physicists to the «profitable» scenarios, which would have resulted from the discovery of neutrino sources, in this case confirms the impression that social constructivism does not supply a satisfactory explanation for the actual work of the researchers (in the field of astroparticle physics).

Other examples concerning the search for violations of the Lorentz Invariance and the search for evidence of dark matter due to the number of positrons and electrons in the cosmic radiation lead to the same result, even if these cases are not so favourable for a constructivist development because the relevant data are obtained by different collaborations.

In summary, it can be said that developments in the field of astroparticle physics (at least insofar as the examples are representative for the research in this discipline) can be understood far better from a realistic than from a constructivist perspective.

3. *Future Constructivist and Realistic Developments in the Field of Astroparticle Physics*

If astroparticle physics were a largely closed discipline, the previous section could be considered as the final conclusion of this study: the dynamics of research in astroparticle physics should therefore certainly be called realistic.

As this area of research is currently undergoing a very active growth phase, the future development of the discipline might call for a revision of this provisional decision in favour of realism (or can alternatively help to support it further). For the social factors that might induce constructivist dynamics of research still remain, like the search for career opportunities, the need to justify significant funding sums, or raising additional funds for subsequent projects, or the pressure of intellectual trends, the expectation of the theorists, etc.

In the previous chapters we have discussed three main scenarios, which imply very profitable prospects for the experimental physicists within the field of astroparticle physics: (1.) the discovery of phenomena that provide an empirical basis for the theoretical program of the quantum gravitation; (2.) the discovery of traces of dark matter; and (3.) the discovery of astrophysical neutrino sources.

The occurrence of the first scenario would provide a subsequent empirical justification for the theoretical search for a quantization of gravity ongoing for more than half a century. In this case, it is conceivable that more information from astroparticle physics would be needed for the construction of the theory of the quantum gravitation. This way theory and experiment would generate a dynamic of mutual fertilization for the benefit of both sides.

The second scenario would clearly confirm the standard model of cosmology, and possibly establish the astroparticle physics even as a key discipline for overcoming the current standard model of particle physics.

The third scenario would open a new window in astrophysics. Many objects could then be seen in a new way that would allow (because of the weak interaction of neutrinos) a very deep look inside the sources.

Since theorists have confidence in the substantial correctness of these scenarios, from a constructivist point of view one should expect that all these opportunities are exploited (or at least some of them). Therefore, it should become increasingly clear in the course of the next years that there is a tendency among astroparticle physicists of getting us closer to the desired goals. That is, negative results should always be put into question; alternative (and more conventional) explanations of the positive results should be ignored, etc.

From a realistic point of view there is no fundamental objection to any of the three scenarios: there could well be quantum gravity effects detectable in the field of astroparticle physics; there could also be products of the annihilation or decay of some types of dark matter, identifiable as astroparticles; and it may well be the case that enough neutrinos are produced in active galaxies or in the supernova explosions (or other sources of astroparticles) to prove the existence of such sources with the help of the neutrino detectors which are currently being constructed.

However, the two first scenarios can be accepted from a realistic viewpoint only if it is convincingly guaranteed that the data given as evidence for them, cannot be naturally explained with the help of already established physics. And the third scenario can only be accepted if all steps of the data analysis have been very carefully examined, among other things, with the help of cross checks, because of the very unfavourable signal/background rate in the case of neutrino experiments. Ideally one should even wait until at least two independent collaborations using different neutrino-detectors achieve the same results.

Consequently, the differences between the realistic and the social constructivist expectations regarding the development of this research field in the next years are clear. So if, for example, the specialists quickly reach an agreement about the fact that the photons from the flares of active galaxies show quantum-gravity-effects, without having previously clarified the details of the dynamics of such flares in a convincing way this should be taken as evidence for a constructivist development. It should also be regarded as a constructivist development when, for example, dark matter is hastily identified as the cause for a surplus of cosmic particles, without exploring first more conventional alternatives, such as pulsars, supernovae, Wolf-Rayet stars or other known sources. What both developments would have in common is the suppression of alternatives resulting from already established physics –suppression unacceptable to a realist–.

A failure of e.g. the IceCube experiment, on the other hand, could hardly be explained from a constructivist point of view. As in the case of the IceCube Collaboration all factors favouring social constructivist dynamics of physics can be found. That is: 1) it is a large collaboration funded with a lot of money; 2) a positive result would mean rosy perspectives for many careers; 3) many new jobs and subsequent investigations could be applied with good prospects for approval; 4) no further checks of the results outside the IceCube collaboration are possible at present; 5) the signal/background rate is so unfavourable – 1/1000000 or even worse– that a highly complicated process of data analysis must be used; 6) the theorists are largely convinced that the effects searched for are real etc. If under such circumstances no construction of the desired phenomena takes place, one should seriously ask whether construction plays a role in the dynamics of physics at all.

Developments as just described can occur in astroparticle physics at any time, or at least as long as the current pace of research in this discipline continues.

Therefore, the verdict of the previous section in favour of realism can only be seen as preliminary. And it is even conceivable that at the time of the publication of this manuscript the situation is totally different. Thus, it seems that the results of this study, as well as currently all results in astroparticle physics, have a very short validity period. Classically minded philosophers may find this prospect to be like a nightmare, for philosophers have always aspired to achieve a secure long-lasting knowledge. The results of this study can obviously not satisfy such an aspiration. But I still hope that they can be understood as a useful contribution to the debate between realism and social constructivism.

4. Realism or sheer Conservatism? About the interpretation of the Examples Examined in this Study

As conclusion of our investigation into philosophy of science it was claimed in the 2nd section of this chapter that dynamics of research in the field of astroparticle physics in a natural way suggest a realistic understanding of the driving forces of research activity, whereas many of the details of this dynamic are hardly understandable from a constructivist perspective. This claim is based not least on the analysis of some examples of actual practice of astroparticle physicists in those cases where the discovery or construction of certain phenomena would open up «profitable» scenarios:

Neither the prospect of linking the theoretical developments in the field of quantum gravity to an empirical field of application nor the prospect of opening a new observational window on astrophysics led to a discovery-fever on the part of the astroparticle physicists or to a rash confirmation of the «profitable» phenomena. Quite the opposite: to date increasingly severe upper limits for the size of the desired phenomena have been set. Beyond that, the announcements of experimental results, which can be interpreted as evidence for the existence of such phenomena, have triggered no avalanche of further confirmations[235] so far. Instead, these results were meticulously re-examined, and rejected.

But now we can ask ourselves what these facts actually mean. In presenting these examples I have always taken (more or less explicitly) the attitude of the experimentalist as an indication that the development, expected from a constructivist perspective for the field of astroparticle physics, does not take place. But is this verdict justified? Could we not just simply say that the crucial factors behind the rejection of some promising hypotheses are also extra scientific, social factors?

Let us think about the fear of the involved physicists and the whole collaborations to lose reputation when they announce results that will later be rejected as incorrect by the international community of specialists. Let us also consider the difficulties in applying for new projects and funding that scientists might have to face after a false discovery. Don't these social factors supply a sufficient reason for the reluctance of experimental physicists in relation to the all too new theoretical proposals?

Moreover, one could argue that even in classical constructivist studies such as «The Golem» by Collins and Pinch some examples –like gravitational waves

---

[235] An example of such by the constructivists described avalanches of further confirmation of a popular hypothesis, after the first signs of its experimental validity are provided, are found e.g. in COLLINS – PINCH (1998) 53:
«Yet after the interpretation of the eclipse observations had come firmly down on the side of Einstein, scientists suddenly began to see confirmation of the red-shift prediction where before they had seen only confusion».

from Weber, or cold fusion– are presented where new ideas are rejected. And such examples are interpreted in a constructivist manner!

Why should we thus present the cases explained here as evidence against the constructivist view on dynamics of research? Would it not be reasonable to claim that such cases merely show that the experimental physicists are very conservative scientists, reluctant to accept the new models?

Now we will try to respond to these objections.

First of all, we must keep in mind the true object of our investigation: For there is certainly a «popular» understanding of social constructivism, which cannot be tested and challenged by any examples. One might summarize this understanding in one sentence: no matter what happens in a research line, we can find extra-scientific reasons that explain the result.

In my opinion, such an approach is undoubtedly legitimate but only represents a caricature of the real social constructivism of Pickering, Collins and Pinch. Because an account compatible with all results cannot make a differentiated statement about dynamics of physics: it is just an empty metaphysical preconception without consequences for our understanding of the development of research.

But the social constructivism proposed in works such as «Constructing Quarks» is much more than that. It is a perspective according to which certain developments are more likely to happen than others in certain research situations. Therefore, Pickering at the end of his study could maintain the following:

> «An exploration of the structure of scientific judgment has therefore been a central preoccupation of this account. I have sought to analyse why scientists should choose to accept the existence of one phenomenon rather than another [...]. I have looked to the relation between particular choices and the contexts in which they were made.»[236]

Let us just consider e.g. the classic example of the confirmation of the existence of neutral weak currents in the abovementioned book. What Pickering affirms at this point is that under the circumstances described by him nothing else than the «discovery» of these currents was to be expected. But this makes it even more apparent, that not every development of a research area is equally corresponding with a constructivist point of view.

This understanding of social constructivism, represented by the most important studies of this school, is the subject of our investigation. That is why I did not single out some random examples from astroparticle physics in the previous chapters, in order to subsequently identify social components for each example which could serve as *ad hoc* explanation for the development that took

---

[236] PICKERING (1984) 405.

place. Instead, I have tried to identify situations that would have developed differently in each case if research dynamics are driven in a rather constructivist and not realistic way.

Against this background, let us again consider the examples mentioned in the previous sections. In my opinion, the critics of the proposed variant of these cases can at least concede that such examples show that the experimental physicist (at least in the field of astroparticle physics) is characterised by a very conservative attitude in relation to the theoretical conceptions, concepts and models. They cling to their proven theoretical tools and face all new hypotheses sceptically at first.

However, it is a fact that in the course of time experimental physicists have repeatedly accepted new hypotheses, have completed old models and have even completely given up old theoretical conceptions. So, how can this be explained?

At this point the realist can offer a pretty simple explanation: it is the pressure of the data collected and evaluated by different teams with different instruments and analysis methods, which leads to a change of opinion among experimental physicists. When the analysis of the data with established methods produces surprising results and several collaborations come to the same results based on different data and analyses, it is not just a coincidence: a new phenomenon has been discovered. And when all conventional attempts to produce an explanation of this phenomenon fail, new theoretical ideas can gain general acceptance. The whole dynamic is driven by the data – *bottom-up*–. Therefore, the phenomena are to be seen as real discoveries and to be interpreted as part of a realistic description of the physical world.

A constructivist must, however, deny this realistic explanation: the reasons for a change of opinion of the initially conservative experimental physicists cannot be found in nature. Because the data are always interpretable in many different ways and they do not force us to recognize any real structures in the world. So how can a constructivist justify the change of opinion of the experimental physicists? Obviously, the driving forces of the physicists cannot be interpreted in terms of a *bottom-up* determination. They must either be *top-down* – that is, from the theoretical hypotheses which are being examined and which fascinate experimental physicists– or they must come from outside –i.e., from extra scientific factors such as the social pressure to justify the received financial support–. The fascination of certain theoretical ideas and/or extra-scientific factors must consequently cause the change of mind of the experimentalists. And the resulting new phenomena are to be understood as constructions which do not possess genuine descriptive value.

However, this implies that certain constructions should (probably) be made despite some opposing factors –such as the aforementioned fear of loss of scientific reputation–,when researchers are in a situation in which the theoretical and the extra-scientific incentives are large enough. For this opposed factors are always present and new ideas nevertheless become accepted time and again. Crucial for such changes must be the theoretical and the extra-scientific incentives which cause experimental physicists to overcome their conservative attitude: in particular the theoretical scenarios developed by an eminent group of scientists –and for which the expected empirical confirmation has yet to be gained– and the consideration of advantages for individual careers and for the further development of collaborations that are associated with the confirmation of such theoretical scenarios.

These considerations can be confirmed by looking at the most important constructivist case studies (within the field of physics). After a detailed description of the –in principle very doubtful– experimental «evidence» supporting the validity of the general theory of relativity, Collins and Pinch e.g. come to the following result:

> «Because relativity was strong, it seemed the natural template through which to interpret the 1919 observations. Because these observations then supported relativity further, the template was still more constraining when it came to dealing with Miller's 1925 observations.»[237]

The confidence in the new theory thus was the crucial factor for the whole development which produced the experimental basis for this theory. And Pickering also recognizes the confidence in the gauge theory of electroweak interactions as the decisive factor for the construction of the confirming phenomena of this theory (the neutral weak currents):

> «Once it had been shown to be renormalisable, gauge theory ceased to be a minority pursuit and became a major theoretical industry. [The gauge theory of the electroweak interaction] predicted the existence of new phenomena in neutrino experiments and one such phenomenon –the weak neutral current– was reported from the Gargamelle neutrino experiment at CERN in 1973 and confirmed at Fermilab the following year. Thus the gauge theory proposed by Yang and Mills in 1954 finally made contact with experiment nineteen years later. The repercussions of this confluence of theory and experiment were enormous, and the concluding section of this chapter therefore explores the experimental discovery of the neutral current in some detail, emphasising the changes in experimental procedure which were an integral part of the production of the new phenomenon.»[238]

The fascinating theoretical hypothesis led to a completely new evaluation of the data of the neutrino experiments and to the introduction of a new experimental way of working in this field, which produced the desired phenomena:

---

[237] COLLINS – PINCH(1998) 52.
[238] PICKERING, A., *Constructing Quarks* 159-160.

«[…] the experimenters had only begun to reconsider their interpretative procedures when „theorists all over the world really started screaming for neutral currents".

The neutrino experimenters did reappraise their interpretative procedure in the early 1970s and […] they succeeded in finding a new set of practices which made the neutral current manifest. The new procedures remained pragmatic and were, in principle, as open to question as the earlier ones. But, like the earlier ones, the new procedures were sustained in the 1970s within a symbiosis of theory and experiment. In adopting the new procedures, the neutrino experimenters effectively got something for nothing.»[239]

On the other hand we find among the constructivist examples of failed discoveries the works of single researchers (or small teams) trying to prove exotic hypotheses[240] –or trying to establish an unsafe method of work[241]; a method regarded by the most experimental physicists as not very promising –. In all these cases the work of the «discoverers» collides with the generally accepted theoretical ideas and expectations (and/or with established procedures of the experimental physicists). Moreover, none of these examples concerns the works of large collaborations where the extra-scientific economic factors appear to be of particular relevance.

Under these circumstances it is not further amazing –from a constructivist perspective– that the new proposals cannot become generally accepted. For many tangible interests stand against them. But if the future of a prestigious line of theoretical research depends on the sought discoveries and these discoveries have been made by large collaborations it is to be expected from a constructivist viewpoint that these discoveries will (probably) take place.

And this is the reason why the examples presented here speak against the constructivist view on dynamics of research. For in these examples all requirements are met, which should lead to the construction of certain phenomena, and there, nevertheless, is no evidence for such constructions at present. When this continues in future it can be said that neither extra-scientific interests nor theoretical trends are the determining factors for the development of astroparticle physics.

---

[239] Ibidem 194.
[240] As e.g. the case of the cool fusion. See COLLINS – PINCH (1998), chap.3.
[241] As e.g. the case of the gravitation waves of Weber. See COLLINS – PINCH (1998), chap.2.

5. About the Scope of the Results of this Study

Let us assume that the realistic interpretation of dynamics of research in the field of astrophysics advocated in this study is correct. What are the consequences?
The answer to this question depends on to what extent one is entitled to assume that astroparticle physics represents a typical case of the *modus operandi* of experimental physics. In my opinion there is sufficient evidence to say that astroparticle physics has no special status in experimental physics:

The articles published in this field found from the beginning their place in the established journals for astrophysics without any systematic conflicts between editors and astroparticle physicists due to methods or content. There is a lively exchange of research personnel between these and other disciplines like astrophysics and particle physics. The engineers and other technical professionals involved in the production and maintenance of the instrumentation in astroparticle physics use the same techniques as in the other disciplines in which they are working. etc.

In addition, some key aspects of research dynamics examined in this study, such as the independence of the work of astroparticle physicists from the new theoretical proposals have already been established in other areas of experimental physics. For example, in the book *Particle Metaphysics* of Brigitte Falkenburg one finds a number of detailed case examples taken from the history of particle physics, which make it apparent that the discoveries in particle physics were made with the help of measurement theories and theories of the measuring instruments, which were clearly independent from the still not secured theoretical proposals in each phase of the development [242]. Falkenburg's warning follows:

> «[...] the authority of a theoretician is *not* an experimenter's authority. Measurement and data analysis are theory-laden, but the theories involved must *not* be identical with the theories under test. The experimenter's job is to keep the theories under test and the theories of measurement or data analysis strictly apart. If they cannot be separated, the measurement goes astray and the experimental results are not regarded as reliable.»[243].

In addition to this independence of the experiments from theoretical proposals the strict *constraints* for possible interpretations of the raw data must also be considered, resulting from the work methods used by particle physicists:

> «[...] in the face of the very complex technical devices and theoretical background knowledge involved [...], physicists always implement as many kinds of redundant measurement methods and internal consistency checks of their

---

[242] See FALKENBURG (2007) chap. 3–5.
[243] FALKENBURG (2007) 55.

data as they can. In current particle physics this redundancy is built into the very design of any experiment. There are independent measurements of the detector properties (including the calibration of the energy scale covered by a detector) and there are as many internal consistency checks of the data as possible. In addition, no experimental result is taken for granted which has not been confirmed by *another* experiment based on *independent* experimental methods»[244].

These words will surely remind the reader of the outline of the working methods of the collaborations in astroparticle physics, summarized in the previous chapters of this book. It is basically one and the same working method. Therefore, it is not surprising that the study of the development of particle physics caused Falkenburg to express criticism about the ductility of the phenomena and the data defended by the social constructivism:

«If theory and measurement are symbiotically interwoven, there are no contingent data. According to Pickering's book *Constructing Quarks*, the data of high energy physics are socio-theoretical construals that depend on school formation within the scientific community. However, it was shown here that the observational basis of particle physics has a layered structure which is built up extremely carefully. In each layer there is a new interplay of observation, measurement and theory. But in each layer only independent measurement methods are considered to be reliable, as long as a theory has not yet been confirmed. The symbiosis of theory and experiment is the result and not the starting point of this development.»[245].

The compliance of the results of the previous chapters with the conclusions Falkenburg obtained in her analysis of the development of particle physics, along with the other above-mentioned signs of «normality» of the astroparticle physics, allow us to state that the results of our study do not represent the outcome of some special features of the methods of astroparticle physics. The claim becomes rather plausible that physics (or at least a large part of experimental physics) develop according to realistic (and not constructivist) dynamics.

However, this result does not imply at all that the dynamics of research in physics cannot become constructivist in the future. At the moment it is not, primarily because the vast majority of physicists are still being engaged (more or less unwittingly) in the Galilean approach modern physics originated from. This approach and its underlying theological roots were mentioned in the first chapter of this study. But as we saw in the first chapter the philosophical debates about realism have a long-lasting effect on the work of physicists. So it cannot be ruled

---

[244] Ibidem 187.
[245] Ibidem 122-123.

out that −when the actual pressure from humanists favouring social constructivism continues− this philosophical framework will eventually influence the way in which physicists understand their work. In other words: the physicists might become constructivists and a constructivist way of doing physics could emerge. This change would have profound consequences, as it implies, among other things, that physicists would have stopped believing in the mathematical rationality of the world and tend to abandon the great ideals of beauty and simplicity of mathematical theories. That would in my opinion lead to something similar to the Ptolemaic Astronomy; a sort of «saving-the-phenomena-science» very different from modern physics. But this is not the place to develop these ideas which deserve their own study.

6. *What Follows when Social Constructivist Analysis of Dynamics of Scientific Research is not Correct?*

Supposing that the arguments presented here are essentially right and that consequently, the social constructivist analysis of dynamics of physical research is not correct, what conclusions should be drawn from it?

The immediate consequence of this result is the confirmation of the thesis that physics describe the reality in which we live. Without doubt, it describes this reality only partially and the description is limited in many ways. But it still gives us an image of nature, containing relevant information about its way of being. But if this is the case, then the following assertion Pickering's −which has been twice quoted in this study due to its central importance− cannot be maintained any more:

> «[…] there is no obligation upon anyone framing a view of the world to take account of what twentieth-century science has to say. […] World-views are cultural products; there is no need to be intimidated by them.»[246]

That is, the results of current physics cannot be regarded as mere cultural constructions without real cognitive value. And since physics obtains a genuine knowledge of the world the results must be taken into account in constructing worldviews that are appropriated for our time. In other words: no contemporary philosophy can aspire to be considered credible if it does not take seriously what physics has to say about the world.

Two important consequences result from this situation: first of all, we must take the thesis very seriously that the world is a reality permeated by mathematical rationality, which at least to some extent determines the «truth» in an objective sense. And secondly, we must accept that the real mathematical

---

[246] PICKERING (1984) 413-414.

structure (or the class of mathematical structures that can be used to characterize the nature) and the resulting objective truth about the world may be incompatible with some ideas of contemporary humanistic and post-modern schools of thought. To get to the heart of it: it must be accepted that our creative freedom is limited by the true rational way of being of the world.

When we now compare this result with some passages from the introduction and the first chapter of this study, where the reasons for the present discontent of the humanists with the natural sciences are described, we can assume that this discontent will be going on in future. Because the social constructivist way to neutralize the «dangers» of the scientific worldview is not convincing and thus these «dangers» –from the viewpoint of some contemporary humanists– will presumably accompany us for a long time.

If we are however ready to accept the fact that the results of physics can not simply be ignored by our search for a general interpretative framework, which makes the things of the world understandable, and gives us orientation in the world, then do not arise for us only limits. Instead, a fascinating puzzle opens. In the words of Polkinghorne:

> «It is scarcely surprising that we can understand the world in the everyday way that is obviously necessary for our survival within it. Yet the development of modern science has shown that human capacity for understanding far exceeds anything that could reasonably be considered as simply an evolutionary necessity, or a happy spin-off from it. We can penetrate the secrets of the subatomic realm of quarks and gluons, and we can make maps of the vast structure of cosmic curved spacetime, both regimes of no direct practical impact upon us, and both exhibiting properties that are counterintuitive in relation to our ordinary habits of thought.
>
> It has also turned out that it is mathematics that is the key to unlocking these scientific secrets. In fundamental physics it is an actual technique of discovery to look for equations that have the unmistakable character of mathematical beauty. […]
>
> When we use abstract mathematics in this way as a guide to physical discovery, something very odd is happening. After all, mathematics is pure thought and what could it be that links that thought to the structure of the physical world around us? […] Eugene Wigner […] once called this "the unreasonable effectiveness of mathematics." He also said it was a gift that we neither deserved nor understood.
>
> Well, I would like to understand it.»[247]

---

[247] POLKINGHORNE (2003) 38-39.

This mystery, which Einstein once called «the eternal mystery of the world» lies before us. It challenges us. But to perceive it, we have to learn to understand the results of physics as more than just a collection of mere cultural constructions.

Bibliography

ABASI, R. et al. (2009a), «Search for high-energy muon neutrinos from the "naked-eye" GRB 080319B with the IceCube neutrino telescope». In: arXiv: 0902.0131.

ABASI, R. et al. (2009b), «Search for muon neutrinos from Gamma-Ray Bursts with the IceCube neutrino telescope». In: arXiv: 0907.2227.

ABDO, A. et al. (2009), *Fermi observations of TeV-selected AGN*. In: ArXiv: 0910.4881

ABDO, A. et al. (2009b), « Testing Einstein's special relativity with Fermi's short hard gamma-ray burst GRB090510» *Nature* (2009). In: arXiv: 0908.1832

ABDO, A. et al (2009c), «Measurement of the cosmic ray $e^+ + e^-$ spectrum from 20 GeV to 1 TeV with the Fermi Large Area Telescope». In: arXiv: 0905.0025

AHRENS, J et al (2004), «Muon Track reconstruction and data selection techniques in AMANDA». In: arXiv: astro-ph/0407044.

AHARONIAN, F. A et al. (2006), «3.9 day orbital modulation in the TeV γ-ray flux and spectrum

from the X-ray binary LS 5039» *Astron. Astrophys*. 460 (2006) 743-49.

AHARONIAN, F. A et al. (2007a), «Detection of extended very-high-energy γ-ray emission towards the young stellar cluster Westerlund 2». In ArXiv:astro-ph/0703427

AHARONIAN F. A, et al. (2007b), «HESS VHE gamma-ray sources without identified counterparts». In: ArXiv:0712.1173

AHARONIAN, F. et al. (2008), «Limits on an Energy Dependence of the Speed of Light from a Flare of the Active Galaxy PKS 2155-304» *Phys. Rev. Lett*. 101 (2008) 170402.

AHARONIAN, F. et al. (2009), «Probing the ATIC peak in the cosmic-ray electron spectrum with HESS». In: arXiv: 0905.0105

ALBERT, J. et al. (2007), «Very High Energy Gamma-ray Radiation from the Stellar-mass Black Hole Cygnus X-1» *Ap.J. Letters* 665 (2007) L51-54.

ALBERT, J. et al. (2008), «Very-high-energy gamma rays from a distant quasar: How transparent is the universe?» *Science* vol.320 no.5884 (2008) 1752-1754.

ALBERT, J. et al. (2008b), «Probing quantum gravity using photons from a flare of the active galactic nucleus Markarian 501 observed by the MAGIC telescope (2007)» *Phys. Rev. Lett* B668 (2008) 253.

ALIU, E. et al. (2008), «Upper Limits on the VHE gamma-ray Emission from the Willman 1 Satellite Galaxy with the MAGIC telescope». In: http://www.arxiv.org/abs/0810.3561.

ALVAREZ-MUÑIZ, J. – MÉSZÁROS, P. (2004), «High energy neutrinos from radio-quiet active galactic nuclei». Phys. Rev. D 70 (12), 123001.

AMELINO-CAMELIA, G. et al. (1998), «Tests of quantum gravity from observations of gamma-ray bursts» *Nature* 393 (1998) 763.

ANDERHUB, H. et al. (2009), «A novel camera type for very high energy gamma-ray astronomy based on Geiger-mode avalanche photodiodes». In: arXiv:0911.4920.

ATWOOD, F. et al. (2009), «The large area telescope on the Fermi gamma-ray space telescope mission». In: arXiv:0902.1089

BACKES, M. (2008*), Langzeitbeobachtung von TeV Blazaren* (Diploma thesis, Dortmund 2008).

BARNES, B. (1991), «How not to do the sociology of knowledge». *Rethinking Objectivity part 1. Special Issue of Annals of Schlolarship* 8 (3-4) 321-335.

BARTKO, H. (2005), «Prototype test of a multiplexed fiber-optic ultra fast FADC Data acquisition system for the MAGIC telescope». Internal Report, MAGIC 2005, MAGIC-TDAS 05-01.

BECKER, J. (2007), «High-energy neutrinos in the context of multimessenger astrophysics». In: ArXiv: 0710.1557

BECKER, J. et al. (2007), «Astrophysical implications of high energy neutrino limits». In: ArXiv: astro-ph/0607427.

BERNARDINI, E. (2009), «The hunt for cosmic neutrino sources with IceCube». In: arXiv:0901.1049.

BIERMANN, P. L. et al. (2009), «Cosmic ray positrons and electrons». In: arXiv: 0903:4048

BIRD, A. (2004), «Thomas Kuhn». In: Stanford Encyclopedia of Philosophy: http://plato.stanford.edu/entries/thomas-kuhn/

BLANCH, O. (2001), *How to use the Camera simulation program.* (Internal Report MAGIC-TDAS 01-04).

BLOBEL, V. (1985), «Unfolding methods in high-energy physics experiments». In: *Proceedings of the 1984 CERN School of Computing.* No. CERN 85-09. (Geneva 1985, CERN) 88.

BLOBEL, V. (1996), «The RUN Manual – Regularized unfolding for high-energy physics experiments». Technical note TN 361, OPAL.

BOYD, R., (2008), *An introduction to nuclear astrophysics* (Chicago, University of Chicago Press).

BRETZ, T. (2006), *Observations of the Active Galactic Nucleus 1 ES 1218+304 with the MAGIC-telescope* (Dissertation at the Universität Würzburg).

BRITZGER, D. (2009), Studies of the Influence of Moonlight on Observations with the MAGIC Telescope (Diplomarbeit, Universität München, June 2009).

BROWN, H. (1993), *Perception, Theory, and Commitment. The New Philosophy of Science* (Chicago 1993, The University of Chicago Press).

BURTT, E. A. (2003), *The Metaphysical Foundations of Modern Science* (New York 2003, Courier Dover Publications).

CARRIER, M., *Nikolaus Kopernikus* (München, Beck 2001).

CASOLINO, M. et al. (2009), «The Pamela cosmic ray space observatory: detector, objectives and first results». In: arXiv:0904.4692

CECCHINI, S. et al. (2009), «Measurement of cosmic ray elemental composition from the CAKE balloon experiment». In: ArXiv 0911.3500

CHALMERS, A. (2005), «Atomism from the 17th to the 20th century». In: Stanford Encyclopedia of Philosophy: http://plato.stanford.edu/entries/atomism-modern/

CHANG, J. et al. (2008), «An excess of cosmic ray electrons at energies of 300–800 GeV» *Nature* 456 (2008) 362-365.

CIRKEL-BARTELT, V. (2008), «History of Astroparticle Physics and its Components» *Living Rev. Relativity* 11 (2008) 2.

CLEVERMANN, F. (2009), *Monte Carlo Simulationen für das IceCube- Projekt und zur Untersuchung der Atmosphäre* (Diploma thesis, Technische Universität Dortmund 2009).

COLLINS, H. (2009), «We Cannot Live by Scepticism Alone» *Nature* 458 (2009) 30-31.

COLLINS, H. (2011), *Gravity's Ghost* (Chicago 2011, The University of Chicago Press).

COLLINS, H. - PINCH, T. (1998), *The Golem. What you should know about science* (Cambridge 1998, Cambridge University Press).

COPERNICUS, N. (1543), *De revolutionibus orbium coelestium*. Available at: http://www.webexhibits.org/calendars/year-text-Copernicus.html

CORTINA, J. (2004), «Status and first results of the MAGIC Telescope». In: http://publications.mppmu.mpg.de/2004/MPP-2004-138/FullText.pdf

DEVITT, M. (1991), «Irrwege der Realismusdebatte». In: H. J. SANDKÜHLER und D. PÄTZOLD (eds.), *Die Wirklichkeit der Wissenschaft. Probleme des Realismus* (Hamburg 1991, Meiner) 113-137.

DIRAC, P.A.M. (1963), «The Evolution of the Physicist's Picture of Nature» *Scientific American* (Mai 1963). Available at: http://www.scientificamerican.com/blog/post.cfm?id=the-evolution-of-the-physicists--pi-2010-06-25

DOERT, M. (2009), *Automated production of standardized and tailor-made Monte Carlo simulations for the MAGIC telescopes* (Diploma thesis, Technische Universität Dortmund 2009).

DREYER, J. (2005), *Hard- und Softwareentwicklung im Rahmen der Experimente AMANDA und IceCube* (Diploma thesis, Technische Universität Dortmund 2005).

DUHEM, P. (1991), *The Aim and Structure of Physical Theory* (Princeton 1991, Princeton University Press).

DUHEM, P. (2002), *Mixture and Chemical Combination and Related Essays* (Dordrecht 2002, Kluwer).

EICHMANN, B. (2009), *Synchroton- und Röntgenvariabilitäten von Blazaren* (Diploma thesis, Universität Bochum).

ERRANDO TRIAS, M. (2009), *Discovery of very high energy gamma-ray emission from 3C 279 and 3C 66A/B with the MAGIC telescope* (Doctoral thesis. Universitat Autònoma de Barcelona) 95.

FALKENBURG, B. (2007) *Particle Metaphysics. A Critical Account of Subatomic Reality* (Heidelberg 2007, Springer).

FALKENBURG, B. (forthcoming), «What are the Phenomena of Physics?». To appear in *Synthese*.

FALKENBURG, B. - RHODE, W. (2007), «Astroteilchenphysik: Die Brücke zwischen Mikro- und Makrokosmos». In: B. FALKENBURG (ed.) (2007), *Natur -*

Technik - Kultur. Philosophie im interdisziplinären Dialog. (Paderborn 2007, Mentis) 57-90.

FRANKLIN, A. (1986), *The Neglect of Experiment*, (Cambridge, 1986, Cambridge University Press).

FRANKLIN, A. (1999), *Can That Be Right? Essays on Experiment, Evidence, and Science* (Dordrecht 1999, Kluwer).

FRANKLIN, A. (2002), *Selectivity and Discord* (Pittsburgh, University of Pittsburgh Press).

FRANKLIN, A. (2009), «Experiments in Physics». In: Stanford Encyclopedia of Philosophy: http://plato.stanford.edu/entries/physics-experiment/

GALISON, P. (1987), *How Experiments End* (Chicago 1987, University of Chicago Press).

GAUG, M. (2006), *Calibration of the MAGIC Telescope and Observation of Gamma Ray Bursts* (Ph. D. Thesis Universitat Autònoma de Barcelona).

GOEBEL, F. et al. (2007), «Upgrade of the MAGIC telescope with a Multiplexed Fiber-Optic 2 GSamples/s FADC Data acquisition system». In: arXiv 0709.2363

HACKING, I. (1999), *The Social Construction of What?* (Harvard 1999, Harvard University Press).

HALZEN, F. (1996), «Status of the AMANDA south pole neutrino detector». In: arXiv:hep-ex/9611014.

HALZEN, F. – HOOPER, D. (2009), «The indirect search for Dark Matter with IceCube». In: arXiv:0910.4513

HELLER, M. (2003), *Creative Tension* (Philadelphia 2003, Templeton Foundation Press).

HINTON, J.A. - HOFMANN, W. (2009), «Teraelectronvolt Astronomy» *Annu. Rev. Astrono. Astrophys.* 47 (2009) p.523-65.

KANITSCHEIDER, B. (2002), *Kosmologie: Geschichte und Systematik in philosophischer Perspektive* (Stuttgart 2002, Reclam).

LATOUR, B. (2005), *Reassembling the Social. An Introduction to Actor-Network-Theory* (New York 2005, Oxford University Press).

LI, T. - MA, Y. (1983), «Analysis methods for results in gamma-ray astronomy» *Astrophysical Journal* 272 (1983) 317.

LÓPEZ CORREDOIRA, M. (2009), «What do astrophysics and the profession have in common?» In LÓPEZ CORREDOIRA, M. – CASTRO PERELMAN, C. (2009), *Against the tide* (Boca Raton, Florida 2009, Universal Publishers) 145-177.

MAGIC COLLABORATION (2008), «Observation of pulsed γ-rays above 25 GeV from the Crab Pulsar with MAGIC». In: ArXiv:0809.2998.

MAGIC COLLABORATION (2009), «Contributions to the 31st international cosmic ray conference». In: arXiv: 0907.0843.

MEADE, P. et al. (2009), «Dark Matter Interpretations of the e+ Excesses after FERMI». In: arXiv:0905.0480

MEYER, M. - MASE, K., (2004), «Analysis of the muon events recorded with the MAGIC telescope». In: http://publications.mppmu.mpg.de/2004/MPP-2004-196/FullText.pdf

MORALEJO, A. (2003), *The Reflector simulation program v.0.6.* (Internal Report MAGIC-TDAS 02-11).

MÜNICH, K. (2007), *Messung des atmosphärischen Neutrinospektrums mit dem AMANDA-II Detektor.* (Doctoral thesis, Technische Universität Dortmund 2007).

NELLEN, L. et al. (1993), «Neutrino production through hadronic cascades in AGN accretion disks». In: ArXiv: hep-ph/9211257.

NIESSE, P. – SPIERING, C. – AMANDA KOLLABORATION (2001), «Search for relativistic monopoles with the AMANDA Detector». In: *Verhandlungen der 27. International Cosmic Ray Conference* (Hamburg 2001).

OTTE, A. N. et al. (2007), «Detection of Cherenkov light from air showers with Geiger-APDs». In: arXiv 0712.1592

PADOVANI, P. - URRY, C. M. (1992), «Luminosity functions, relativistic beaming, and unified theories of high-luminosity radio sources»: *Astrophysical Journal* 387 (1992) 449.

PANEQUE CAMARERO, D. (2004), *The MAGIC Telescope: development of new technologies and first observations* (Dr. Thesis, Technische Universität München).

PICKERING, A. (1984), *Constructing Quarks* (Edingburgh 1984, Edingburgh University Press).

PICKERING, A. (1984b), «Against putting the phenomena first: The Discovery of the weak neutral current» *Studies in the History and Philosophy of Science* 15 (1984) 85-117.

PICKERING, A. (1991), «Philosophy naturalized a bit» *Social Studies of Science* 21 (1991) 578.

PICKERING, A. (1995), *The Mangle of Practice* (Chicago 1995, Chicago University Press).

POLKINGHORNE, J. (1983), «Quest for a natural God». *Times Higher Educational Supplement* (10.Jun.1983) 24.

POLKINGHORNE, J. (2003), «Physics and Metaphysics in a Trinitarian Perspective» *Theology and Science*, Vol. 1, No. 1 (2003) 38-39

POLKINGHORNE, J. (2003), «Physics and methaphysics in a trinitarian perspective» *Theology and Science*, Vol. 1, No. 1 (2003).

RIOJA, A. - ORDÓÑEZ, J. (1999), *Teorías del Universo* Vol. 1 (Madrid 1999, Síntesis).

ROBBINS, B. (1996), «Anatomy of a hoax» *Tikkun*, September/October (1996) 58-59.

SCHNEIDER, P. (2006), *Einführung in die extragalaktische Astronomie und Kosmologie* (Berlin 2006, Springer).

SCHUBNELL, M. (2009), «The cosmic-ray positron and electron excess: an experimentalist's point of view». In: arXiv: 0905.0444

SCHULZ, C. (2008) *Novel All-Aluminium Mirrors of the MAGIC Telescope Project and Low Light Level Silicon Photo-Multiplier Sensors for Future Telescopes* (Diploma thesis, Fachhochschule München).

SELLARS, W. (1962), *Science, Perception and Reality* (New York 1962, Humanities Press).

SHAO, L. - XIAO, Z. - MA, B. (2009), «Lorentz violation from cosmological objects with very high energy photon emissions». In: arXiv 0911.2276

SOBCZYŃSKA, D. (2002), *Mmscs from CORSIKA 6.014*. (Internal Report MAGIC-TDAS 02-10).

SOKAL, A. (1996a), «Transgressing the boundaries: Towards a transformative hermeneutics of quantum gravity» *Social Text* 46/47 (1996) 217-252.

SOKAL, A. (1996b), «A physicist experiments with cultural studies» *Lingua Franca* May/June (1996) 62-64.

SOKAL, A. (1998), «Truth, reason, objectivity and the left» *Economic and Political Weekly* 18 April (1998) 913–4.

SOKOLSKY, P. et al. (2005), «Comparison of UHE Composition Measurements by Fly's Eye, HiRes-prototype/MIA and Stereo HiRes Experiments». In: arXiv:astro-ph/0507485

STANEV, T. (2004), *High energy cosmic rays* (Berlin 2004, Springer).

STECKER, F. – SALAMON, M. (1996), «High energy neutrinos from quasars». In: ArXiv: astro-ph/9501064.

STÖCKLER, M. (2003), «Über die vielen Formen des Realismus». In: H. REUTER - B. BRECKLING - A. MITTWOLLEN (eds.) (2003), *Gene, Bits und Ökosysteme* (Frankfurt a. Main 2003, Peter Lang) 235-245.

STRAHLER, E. (2009), «Recent results of point source searches with the IceCube neutrino telescope». In: arXiv:0905.4705.

TESCARO, D. et al. (2007), «Study of the performance and capability of the new ultra fast 2 GSample/s FADC data acquisition system of the MAGIC telescope». In: arXiv 0709.1410

THOM, M. (2009), *Analyse der Quelle 1ES 1959+650 mit MAGIC und die Implementierung eines Webinterfaces für die automatische Monte-Carlo-Production* (Diploma thesis, Dortmund 2009).

TORRETTI, R. (2004), «Andrew Pickering, Constructing Quarks, A sociological History of Particle Physics». In: M. ESPINOZA and R.TORRETTI (2004), *Pensar la ciencia* (Madrid 2004, Tecnos) 231-238.

TORRETTI, R. (2007), *De Eudoxo a Newton: Modelos Matemáticos en la Filosofía Natural* (Santiago de Chile 2007, Ediciones de la Universidad Diego Portales).

VAN ELEWYCK, V. (2007), «Pierre Auger observatory status and results». In: arXiv:0709.1422

VAN FRAASEN, B. (1980), *The Scientific Image* (Oxford 1980, Clarendon Press).

WEEKES, T. et al. (1989), «Observation of TeV gamma rays from the Crab nebula using the atmospheric Cherenkov imaging technique» *The Astrophysical Journal* 342 (1989) 379-395.

WELTEVREDE, P. et al. (2009), «Gamma-ray and radio properties of six pulsars detected by the Fermi large area telescope». In: ArXiv:0911.3063.

WIEDEMANN, A. (2008), *Energy spectra unfolding of atmospheric neutrinos with the IceCube 22-string detector* (Diploma thesis, Technische Universität Dortmund 2008).

WIGMANS, R. (2000), *Calorimetry* (Oxford 2000, Oxford Science Publications)

ZECH, A. (2005), «Studies of Systematic Uncertainties in the HiRes-II Measurement of the UHECR Spectrum». In: arXiv:astro-ph/0507401

Index

A

ABASI 182, 212
ABDO 59, 109, 111, 212
Active Galactic Nuclei 55, 58, 170, 190
AGN 55, 58, 59, 60, 63, 96, 97, 107, 159, 170, 190, 192, 212, 217
AHARONIAN 57, 58, 108, 110, 212
AHRENS 181, 212
air shower ground experiments 67, 79
ALBERT 58, 60, 107, 108, 212, 213
ALIU 157, 213
ALVAREZ-MUÑIZ 193, 213
AMANDA 15, 101, 160, 165, 168, 175, 177, 185, 186, 188, 189, 190, 191, 192, 193, 194, 195, 198, 212, 215, 216, 217
AMANDA KOLLABORATION 217
AMELINO-CAMELIA 107, 213
analysis chain 143, 146, 148, 150, 198
ANDERHUB 133, 213
Antares 85
anti-realism 20, 26, 37
Aristotle 19, 20, 22
artifacts 141, 147, 156
ATIC 75, 109, 110, 212
ATIC peak 109
atmospheric neutrinos 62, 63, 181, 183, 184, 185, 194, 195, 219
ATWOOD 73, 213

B

BACKES 7, 51, 213
Baikal 85
balloon experiments 52, 66, 67, 74, 75, 83
BARNES 194, 213
BARTKO 131, 213
BECKER 52, 60, 61, 62, 193, 213
BERNARDINI 184, 213
Bernlöhr 67
BIERMANN 110, 213
binned analysis 184
BIRD 27, 213
BLANCH 153, 213
BLOBEL 185, 214
Bogen 46, 65
Bohr 24
Boltzmann 24
BOYD 192, 214
Boyle 23
BRETZ 144, 214
BRITZGER 214
BROWN 41, 214
BURTT 37, 214

C

calibration 71, 73, 74, 79, 83, 88, 89, 105, 112, 114, 117, 121, 123, 124, 131, 133, 134, 135, 138, 139, 143, 148, 151, 162, 167, 174, 178, 188, 189, 197, 207
Callippus 19
calorimeter 68, 69, 71, 73, 74, 75, 76, 88
CANGAROO 84
career interests 9, 26, 42
CARRIER 22, 214
CASOLINO 68, 70, 214
CECCHINI 52, 214
CERN 45, 65, 204, 214

CHALMERS 24, 214
CHANG 109, 214
Cherenkov light 66, 81, 83, 85, 116, 118, 121, 126, 132, 133, 142, 169, 175, 181, 182, 186, 188, 217
Cherenkov telescopes 66, 84, 90, 118, 119, 131, 134, 135, 138, 141, 151, 154, 155, 156
CIRKEL-BARTELT 50, 214
cleanup of the background 145
CLEVERMANN 7, 182, 214
Cline 46
collaboration board 165, 166, 167
COLLINS 28, 43, 44, 46, 87, 88, 123, 124, 136, 137, 138, 156, 201, 202, 204, 205, 214, 215
consistency checks 139, 157, 206
COPERNICUS 21, 22, 215
CORSIKA 152, 153, 154, 155, 182, 197, 218
CORTINA 127, 134, 135, 215
cosmic neutrinos 62, 84, 183, 185, 191, 194, 195
cosmic rays 48, 50, 51, 52, 54, 66, 67, 68, 72, 73, 74, 77, 79, 83, 85, 89, 90, 91, 94, 99, 109, 219
cosmology 7, 20, 21, 22, 48, 50, 63, 98, 199
Crab Nebula 56, 131, 134, 135, 138, 139, 140, 157
cross check 127
cuality cuts 147
cultural products 14, 29, 36, 138, 208

D

Dalton 23
dance of agency 33

dark matter 7, 49, 98, 99, 100, 103, 106, 109, 110, 111, 157, 168, 170, 190, 198, 199, 200
data acquisition system 73, 84, 122, 124, 125, 130, 131, 153, 219
data analysis 10, 15, 33, 36, 45, 48, 64, 65, 71, 74, 79, 84, 87, 89, 90, 105, 106, 111, 112, 113, 114, 116, 118, 122, 141, 142, 147, 148, 154, 155, 156, 157, 158, 160, 166, 167, 177, 179, 180, 182, 183, 185, 186, 189, 192, 193, 197, 200, 206
DEVITT 41, 215
DIRAC 37, 38, 40, 215
DOERT 7, 56, 153, 215
DOMs 175, 177, 178, 180, 181, 188
DREYER 7, 174, 175, 188, 215
DUHEM 23, 25, 37, 215

E

EGRET 56, 134, 138
EICHMANN 7, 97, 215
Einstein 24, 109, 201, 210, 212
ERRANDO TRIAS 149, 215
Eudoxus 19, 20
executive committee 166, 167
experimental results 26, 29, 32, 64, 79, 87, 102, 107, 201, 206
experimentalists 29, 31, 32, 33, 36, 44, 47, 65, 87, 88, 89, 102, 103, 107, 109, 110, 111, 194, 203
experimenters' regress 137, 141
extra-scientific social forces 161

F

FALKENBURG 7, 10, 16, 24, 41, 47, 64, 206, 207, 215

Fermi 56, 71, 94, 95, 96, 108, 109, 110, 112, 212, 213, 219
Fermi acceleration 94
FERMI LAT 68, 72, 74
Fermilab 45, 46, 204
fluorescence experiments 67, 77, 78
FRANKLIN 7, 9, 10, 26, 27, 34, 40, 41, 42, 43, 44, 46, 106, 110, 140, 141, 147, 216

G

Galileo 22, 37, 140, 141
GALISON 10, 43, 44, 45, 46, 158, 216
Gamma flares 58
gamma flux 56, 92, 100, 138, 157
gamma-ray bursts 60, 61, 64, 73, 90, 106, 109, 213
gamma-rays 54, 55, 57, 58, 59, 60, 62, 66, 67, 74, 76, 93, 100
Gargamelle 45, 46, 65, 204
Gassendi 23
GAUG 121, 216
GOEBEL 131, 216
gravitational waves 44, 124, 137, 201

H

HACKING 10, 161, 162, 216
hadron events 130, 143, 146, 147, 150
hadronic model 97
HALZEN 100, 191, 216
HELLER 41, 216
Hess 50
HESS 57, 84, 90, 107, 108, 110, 111, 212
high-energy gamma-radiation 92
high-energy particles 10, 50, 79

Hillas Parameters 145, 149
HINTON 55, 216
HiRes 78, 79, 219, 220
hodoscope 75
HOFMANN 55, 216
HOOPER 100, 216

I

IceCube 14, 15, 85, 86, 89, 101, 106, 113, 159, 160, 165, 166, 167, 168, 169, 170, 171, 174, 175, 176, 177, 178, 179, 180, 181, 183, 184, 185, 186, 188, 189, 190, 191, 192, 195, 196, 200, 212, 213, 214, 215, 216, 219
IceCube Detector 15, 168
IceCube Experiment 15, 168
Image Generating Air Cherenkov Telescopes 83
interactions between theories and experiments 50
interests 9, 12, 29, 31, 33, 44, 114, 115, 138, 142, 154, 155, 162, 189, 205
inverse Compton scattering 93

K

Kaluza-Klein particles 110
KANITSCHEIDER 19, 20, 22, 216
Kepler 22, 27, 98, 141
Kuhn 27, 213

L

LATOUR 156, 161, 216
LEDs 121, 174
leptonic model 92, 97
LI 147, 216

LÓPEZ CORREDOIRA 162, 164, 167, 195, 217
Lorentz Invariance 101, 102, 103, 106, 107, 108, 109, 111, 198

M

MA 107, 147, 216, 218
Mach 23, 24, 25, 37
MAGIC 14, 15, 56, 57, 58, 60, 84, 88, 89, 90, 106, 107, 112, 113, 114, 115, 116, 117, 118, 119, 120, 121, 122, 123, 124, 125, 126, 127, 128, 129, 130, 131, 132, 133, 134, 135, 136, 138, 139, 141, 142, 148, 150, 151, 152, 153, 154, 155, 156, 157, 158, 160, 167, 179, 180, 182, 183, 188, 196, 213, 214, 215, 216, 217, 218, 219
MAGIC COLLABORATION 15, 56, 113, 115, 120, 123, 130, 142, 148, 154, 155, 156, 157, 158, 160, 167, 217
MAGIC Telescope 113, 116, 126, 130, 134, 135, 214, 215, 216, 217, 218
MAGIC telescopes 84, 89, 116, 117, 118, 119, 123, 124, 133, 138, 139, 141, 155, 215
MAGIC Telescopes 15, 113, 115, 117, 119, 123, 124, 125, 131, 133, 136, 142, 154, 158
MASE 135, 217
Mayo 46
MEADE 99, 100, 217
MÉSZÁROS 193, 213
MEYER 135, 217
Millikan 25
Monte Carlo Simulations 144, 145, 148, 150, 151, 154, 189
MORALEJO 152, 217

Morpurgo 34
MÜNICH 185, 217
muon background 135, 183
muon rings 135, 147

N

negotiation 26
NELLEN 193, 217
Nemo 85
Nestor 85
neutrino spectrum 62, 63, 192, 193
NIESSE 101, 217

O

open star clusters 58
Open Stars-Clusters 55
order of nature 12, 13, 14, 15, 41, 196
ORDÓÑEZ 218
Osiander 21, 22, 26
OTTE 131, 132, 217

P

PADOVANI 59, 217
PAMELA 68, 69, 70, 71, 73, 74, 109, 112
PANEQUE CAMARERO 128, 134, 217
particle accelerators 45, 62, 80, 91, 154, 158
particle air shower 66
Perrin 25
phenomena 7, 9, 10, 14, 15, 17, 19, 20, 21, 23, 24, 25, 33, 34, 36, 37, 40, 44, 48, 49, 50, 51, 52, 54, 59, 60, 61, 62, 64, 65, 66, 77, 87, 88, 89, 90, 91, 96, 98, 99, 102, 103, 104, 105, 106, 109, 111, 112, 113, 115, 138, 141, 142, 147,

154, 155, 156, 157, 158, 159, 160, 168, 170, 171, 190, 191, 194, 197, 198, 199, 200, 201, 203, 204, 205, 207, 208, 217
photomultiplier 81, 85, 121, 174
physical models 17, 25
physical phenomena 10, 11, 14, 19, 35, 188
physical theories 24, 25, 28, 29, 38, 41, 43
PICKERING 13, 14, 27, 28, 29, 30, 31, 33, 34, 42, 43, 44, 45, 46, 87, 88, 105, 156, 158, 194, 202, 204, 207, 208, 217, 218, 219
Pierre Auger Experiment 80, 81
PINCH 43, 44, 46, 123, 124, 136, 137, 201, 202, 204, 205, 215
Planck 24, 38, 67, 93, 101, 102, 113, 117
Plato 18, 19, 39
PMTs 119, 120, 121, 122, 131, 132, 142, 143, 153, 175
point spread function 125, 126, 127
POLKINGHORNE 28, 40, 43, 209, 218
Pope Paul III 21
production mechanisms of gamma-radiation 92
Pulsar Wind Nebulae 55

Q

quantum gravity 12, 49, 90, 101, 102, 103, 104, 107, 108, 111, 199, 201, 213, 218
quantum theory of gravity 50

R

Random Forest 149, 189

raw data 36, 64, 65, 74, 84, 87, 89, 115, 121, 122, 124, 142, 144, 160, 180, 181, 197, 206
real structure of the world 7, 21, 29, 41, 161
realism 7, 10, 11, 14, 15, 17, 18, 20, 22, 23, 24, 25, 26, 36, 37, 41, 47, 48, 51, 71, 86, 104, 106, 113, 140, 156, 160, 189, 191, 194, 196, 199, 200, 207
RHODE 7, 47, 215
RIOJA 19, 22, 218
ROBBINS 12, 13, 218
Rutherford 24

S

SALAMON 193, 219
satellite data 57, 138
satellite detectors 67, 68, 79, 119
SCHNEIDER 98, 218
SCHUBNELL 109, 218
SCHULZ 126, 127, 218
science wars 9, 12, 161
scientific hypotheses 26
scientific judgments 27, 30, 31
SELLARS 25, 26, 218
SHAO 107, 218
shock acceleration 92, 94
signal/background 48, 106, 146, 177, 179, 180, 200
SOBCZYŃSKA 152, 218
social constructions 9, 11, 12, 13
social constructivism 7, 10, 11, 14, 15, 26, 27, 36, 37, 42, 47, 48, 86, 104, 160, 188, 189, 191, 194, 195, 196, 198, 200, 202, 207, 208
social constructivist explanation 10
social constructivist studies 9, 10, 27, 42
SOKAL 12, 13, 218

SOKOLSKY 78, 219
SPIERING 101, 217
standard model 26, 48, 50, 63, 99, 199
STANEV 52, 219
STECKER 193, 219
stereo Fly's Eye 78
STÖCKLER 17, 219
STRAHLER 183, 184, 219
SuperKamiokande 85
supernova 1987A 62
Supernovae Remnants 55
supersymetric models 99
synchrotron radiation 93, 97

T

TESCARO 131, 219
theoretical models 15, 34, 36, 50, 51, 54, 90, 91, 96, 97, 99, 100, 101, 102, 103, 105, 107, 109, 154, 185, 191, 192, 193, 194, 197
theorists 15, 29, 31, 32, 33, 34, 41, 49, 51, 74, 79, 100, 103, 104, 107, 108, 109, 110, 111, 194, 199, 200, 205
THOM 7, 60, 148, 151, 156, 219
Thomson 25
TIGRE 75, 76, 77
TORRETTI 19, 20, 39, 42, 43, 219
tracker system 68
trigger system 68, 82, 83, 122
truth 12, 13, 21, 31, 40, 208

U

unbinned analysis 184

underground detectors 67
Unidentified Galactic Sources 55, 58
URRY 59, 217

V

VAN ELEWYCK 83, 219
VAN FRAASEN 25, 26, 41, 219
VERITAS 84, 90, 107

W

weak neutral currents 34, 45, 46, 48, 158
Weber 44, 124, 136, 137, 202, 205
WEEKES 134, 139, 140, 219
WELTEVREDE 56, 219
WHIPPLE Telescope 135, 139
WIEDEMANN 181, 219
WIGMANS 88, 220
WIMPs 99, 100, 170, 190, 191, 192, 194
Wolf-Rayet stars 58, 61, 62, 200
Woodward 46, 65
worldview of physics 11, 14, 38

X

XIAO 107, 218
X-ray Binaries 55

Z

ZECH 79, 220

PHILOSOPHIE UND GESCHICHTE DER WISSENSCHAFTEN

Studien und Quellen

In dieser Buchreihe, in der seit 1988 mehr als 70 Bände erschienen sind, werden in der Perspektive des Pluralismus philosophischen Denkens Untersuchungen sowohl zur Theoretischen und Praktischen Philosophie als auch zu deren Geschichte veröffentlicht. In ihrer Gesamtheit sind die Beiträge zugleich ein Plädoyer für ein Philosophieren, das die Nähe zu den Wissenschaften in ihrer historischen und aktuellen Entwicklung sucht.

Band 1 Roland Daniels: Mikrokosmos. Entwurf einer physiologischen Anthropologie. Erstveröffentlichung des Manuskripts von 1850/51, hg. vom Karl-Marx-Haus, Trier. 1988.

Band 2 V.A. Lektorskij: Subjekt – Objekt – Erkenntnis. Grundlegung einer Theorie des Wissens. 1985.

Band 3 Martin Hundt: Geschichte des Bundes der Kommunisten 1836 bis 1852. 1993.

Band 4 Lothar Knatz: Utopie und Wissenschaft im frühen deutschen Sozialismus. Theoriebildung und Wissenschaftsbegriff bei Wilhelm Weitling. 1984.

Band 5 Ferdinando Vidoni: Ignorabimus! Emil du Bois-Reymond und die Debatte über die Grenzen wissenschaftlicher Erkenntnis im 19. Jahrhundert. Mit einem Vorwort von Ludovico Geymonat. 1991.

Band 6 Vesa Oittinen: Spinozistische Dialektik. Die Spinoza-Lektüre des französischen Strukturalismus und Poststrukturalismus. 1994.

Band 7 Lars Lambrecht: Intellektuelle Subjektivität und Gesellschaftsgeschichte. Grundzüge eines Forschungsprogramms zur Biographik und Fallstudie zu F. Nietzsche und F. Mehring. 1985.

Band 8 Frank Unger: Politische Ökonomie und Subjekt der Geschichte. Empirie und Humanismus als Voraussetzung materialistischer Geschichtstheorie. 1985.

Band 9 Bernhard Delfgaauw, Hans Heinz Holz, Lolle Nauta (Hrsg.): Philosophische Rede vom Menschen. Studien zur Anthropologie Helmuth Plessners. 1985.

Band 10 Christfried Tögel (Hrsg.): Struktur und Dynamik wissenschaftlicher Theorien. Beiträge zur Wissenschaftsgeschichte und Wissenschaftstheorie aus der bulgarischen Forschung. 1986.

Band 11 Gerhard Pasternack (Hrsg.): Philosophie und Wissenschaften. Das Problem des Apriorismus. 1987.

Band 12 Gerhard Pasternack (Hrsg.): Philosophie und Wissenschaften. Zum Verhältnis von ontologischen, epistemologischen und methodologischen Voraussetzungen der Einzelwissenschaften. 1990.

Band 13 Wilhelm Weitling: Grundzüge einer allgemeinen Denk- und Sprachlehre. Herausgegeben und eingeleitet von Lothar Knatz. 1991.

Band 14 Hans Jörg Sandkühler (Hrsg.): Geschichtlichkeit der Philosophie. Theorie, Methodologie und Methode der Historiographie der Philosophie. 1991.

Band 15 Rolf-Dieter Vogeler: Engagierte Wissenschaftler. Bernal, Huxley und Co.: Über das Projekt der "Social Relations of Science"-Bewegung. 1992.

Band 16 Volkmar Schöneburg (Hrsg.): Philosophie des Rechts und das Recht der Philosophie. Festschrift für Hermann Klenner. 1992.

Band 17 Axel Horstmann: Antike Theoria und moderne Wissenschaft. August Boeckhs Konzeption der Philologie. 1992.

Band 18	Hans Jörg Sandkühler (Hrsg.): Wirklichkeit und Wissen. Realismus, Antirealismus und Wirklichkeits-Konzeptionen in Philosophie und Wissenschaft. 1992.
Band 19	Georg Quaas: Dialektik als philosophische Theorie und Methode des 'Kapital'. Eine methodologische Untersuchung des ökonomischen Werkes von Karl Marx. 1992.
Band 20	Volker Schürmann: Praxis des Abstrahierens. Naturdialektik als relationsontologischer Monismus. 1993.
Band 21	Theodor Celms: Der phänomenologische Idealismus Husserls und andere Schriften 1928–1943. Herausgegeben von Juris Rozenvalds. 1993.
Band 22	Hans Jörg Sandkühler (Hrsg.): Konstruktion und Realität. Wissenschaftsphilosophische Studien. 1994.
Band 23	Hans Jörg Sandkühler (Hrsg.): Freiheit, Verantwortung und Folgen in der Wissenschaft. 1994.
Band 24	Hyong-Sik Yun: Semiotische Tätigkeitsphilosophie. Interner Realismus in neuer Begründung. 1994.
Band 25	Hans Jörg Sandkühler (Hrsg.): Theorien, Modelle und Tatsachen. Konzepte der Philosophie und der Wissenschaften. 1994.
Band 26	Detlev Söffler: Auf dem Weg zu Kants Theorie der Zeit. Untersuchung zur Genese des Zeitbegriffs in der Philosophie Immanuel Kants. 1994.
Band 27	Hyondok Choe: Gaston Bachelard. Epistemologie. Bibliographie. 1994.
Band 28	Reinhard Mocek: Johann Christian Reil (1759–1813). Das Problem des Übergangs von der Spätaufklärung zur Romantik in Biologie und Medizin in Deutschland. 1995.
Band 29	Detlev Pätzold: Spinoza – Aufklärung – Idealismus. Die Substanz der Moderne. 1995.
Band 30	Martina Plümacher: Identität in Krisen. Selbstverständigungen und Selbstverständnisse der Philosophie in der Bundesrepublik Deutschland nach 1945. 1995.
Band 31	Hans Jörg Sandkühler (Hrsg.): Interaktionen zwischen Philosophie und empirischen Wissenschaften. Philosophie- und Wissenschaftsgeschichte zwischen Francis Bacon und Ernst Cassirer. 1995.
Band 32	Seungwan Han: Marx in epistemischen Kontexten. Eine Dialektik der Philosophie und der 'positiven Wissenschaften'. 1995.
Band 33	Martina Plümacher / Volker Schürmann (Hrsg.): Einheit des Geistes. Probleme ihrer Grundlegung in der Philosophie Ernst Cassirers. 1996.
Band 34	Dagmar Borchers: Der große Graben – Heidegger und die Analytische Philosophie. 1997.
Band 35	Anneliese Griese / Hans Jörg Sandkühler (Hrsg.): Karl Marx – zwischen Philosophie und Naturwissenschaften. 1997.
Band 36	Hans Jörg Sandkühler (Hrsg.): Philosophie und Wissenschaften. Formen und Prozesse ihrer Interaktion. 1997.
Band 37	Hyondok Choe: Ideologie. Eine Geschichte der Entstehung des gesellschaftskritischen Begriffs. 1997.
Band 38	Wolfgang Wildgen: Das kosmische Gedächtnis. Kosmologie, Semiotik und Gedächtnistheorie im Werke Giordano Brunos (1548–1600). 1998.
Band 39	Johann Jakob Wagner: Dictate über Ideal- und Naturphilosophie. Herausgegeben und historisch-kritisch kommentiert von Stefano Palombari. 1998.
Band 40	Hongqing Quan: Das Problem des Selbstbewußtseins, seine Begründung in Kants Philosophie und im Neo-Konfuzianismus. 1998.

Band 41	Roberto Finelli: Mythos und Kritik der Formen. Die Jugend Hegels (1770–1803). 2000.
Band 42	Hans Jörg Sandkühler (Hrsg.): Naturverständnisse, Moral und Recht in der Wissenschaft. Zur Problematik von Tierversuchen. 2000.
Band 43	Joon-kee Hong: Der Subjektbegriff bei Lacan und Althusser. Ein philosophisch-systematischer Versuch zur Rekonstruktion ihrer Theorien. 2000.
Band 44	Wilhelm Büttemeyer / Hans Jörg Sandkühler (Hrsg.): Übersetzung – Sprache und Interpretation. 2000.
Band 45	Hans Jörg Sandkühler (Hrsg.): Selbstrepräsentation in Natur und Kultur. 2000.
Band 46	Sandro Nannini / Hans Jörg Sandkühler (eds.): Naturalism in the Cognitive Sciences and the Philosophy of Mind. 2000.
Band 47	Juichi Matsuyama / Hans Jörg Sandkühler (Hrsg.): Natur, Kunst und Geschichte der Freiheit. Studien zur Philosophie F.W.J. Schellings in Japan. 2000.
Band 48	Jindrich Zeleny: Die dialektische Ontologie. 2001.
Band 49	Christian Westphal: Von der Philosophie zur Physik der Raumzeit. 2002.
Band 50	Francesco Fistetti: Heidegger und die Utopie der Polis. 2002.
Band 51	Reinhard Mocek: Biologie und soziale Befreiung. Zur Geschichte des Biologismus und der *Rassenhygiene* in der Arbeiterbewegung. 2002.
Band 52	Hans Jörg Sandkühler (Hrsg.): Welten in Zeichen – Sprache, Perspektivität, Interpretation. 2002.
Band 53	Giorgio Baratta: Das dialogische Denken Antonio Gramscis. 2003.
Band 54	Silja Freudenberger / Hans Jörg Sandkühler (Hrsg.): Repräsentation, Krise der Repräsentation, Paradigmenwechsel. Ein Forschungsprogramm in Philosophie und Wissenschaften. 2003.
Band 55	Francisco José Soler Gil: Aristoteles in der Quantenwelt. Eine Untersuchung über die Anwendbarkeit des aristotelischen Substanzbegriffes auf die Quantenobjekte. 2003.
Band 56	Paweł Dybel / Hans Jörg Sandkühler (Hrsg.): Der Begriff des Subjekts in der modernen und postmodernen Philosophie. 2004.
Band 57	Hans Jörg Sandkühler / Hong-Bin Lim (eds.): Transculturality – Epistemology, Ethics, and Politics. 2004.
Band 58	Lars Lambrecht / Bettina Lösch / Norman Paech (Hrsg.): Hegemoniale Weltpolitik und Krise des Staates. 2006.
Band 59	Zanghyon Bak: Das Menschen- und Weltbild bei Feuerbach und Marx. Zur Begründung der Solidarität. 2006.
Band 60	Sandro Nannini: Seele, Geist und Körper. Historische Wurzeln und philosophische Grundlagen der Kognitionswissenschaften. Deutsche Bearbeitung von Sibylle Mahrdt. 2006.
Band 61	Jens Pape: Der Spiegel der Vergangenheit. Geschichtswissenschaft zwischen Relativismus und Realismus. 2006.
Band 62	Valerio Meattini: Der Ort des Verstehens. 2007.
Band 63	Sandra Baquedano Jer: Wille zur Phantasie. Versuch das „Nichts" bei Schopenhauer auszuloten. 2007.
Band 64	Hans Jörg Sandkühler (Hrsg.): Menschenwürde. Philosophische, theologische und juristische Analysen. 2007.
Band 65	Sara Dellantonio: Die interne Dimension der Bedeutung. Externalismus, Internalismus und semantische Kompetenz. 2007.

Band 66   Hans Jörg Sandkühler (Hrsg.): Repräsentation und Wissenskulturen. 2007.

Band 67   Mario Trevi / Marco Innamorati: Das Erbe C. G. Jungs. Neue Wege in der dynamischen Psychologie. 2008.

Band 68   Werner Goldschmidt / Bettina Lösch / Jörg Reitzig (Hrsg.): Freiheit, Gleichheit, Solidarität. Beiträge zur Dialektik der Demokratie. 2009.

Band 69   Melanie Hoffmann: Wissenskulturen, Experimentalkulturen und das Problem der Repräsentation. 2009.

Band 70   Mariannina Failla: Hans-Georg Gadamer als Platon-Interpret: Die Musik. 2009.

Band 71   Luigi Pastore (Hrsg.): Subjekt, Objekt, Repräsentation. Zwischen Konstitution und Konstruktion. 2010.

Band 72   Guido Seddone: Stufen des *Wir*. Gemeinschaft als Basis personalen Handelns. 2011.

Band 73   Francisco José Soler Gil: Discovery or Construction? Astroparticle Physics and the Search for Physical Reality. 2012.

www.peterlang.de

www.ingramcontent.com/pod-product-compliance
Ingram Content Group UK Ltd.
Pitfield, Milton Keynes, MK11 3LW, UK
UKHW042003230426
12048UKWH00009B/508